Introduction

Welcome to Collins Bridging GCSE and A Level Maths Student Book. This book helps you to progress smoothly onwards from GCSE Maths with detailed examples and plenty of practice in the key areas needed for success at A Level.

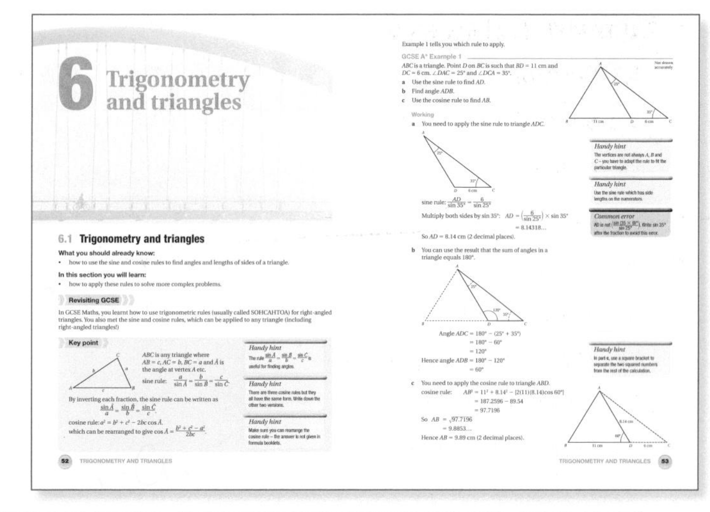
6 Trigonometry and triangles

6.1 Trigonometry and triangles

What you should already know:

- how to use the sine and cosine rules to find angles and lengths of sides of a triangle.

In this section you will learn:

- how to apply these rules to solve more complex problems.

Revisiting GCSE

Key point

Handy hint

Handy hint

Handy hint

Example 1 tells you which rule to apply.

GCSE A* Example 1

Working

Handy hint

Handy hint

Common error

Handy hint

52 TRIGONOMETRY AND TRIANGLES

TRIGONOMETRY AND TRIANGLES 53

Revisiting GCSE

Consolidate your knowledge of difficult GCSE topics with graded worked examples.

Moving on to AS Level

Find out how the strategies you learnt at GCSE are extended and explored with AS Level worked examples.

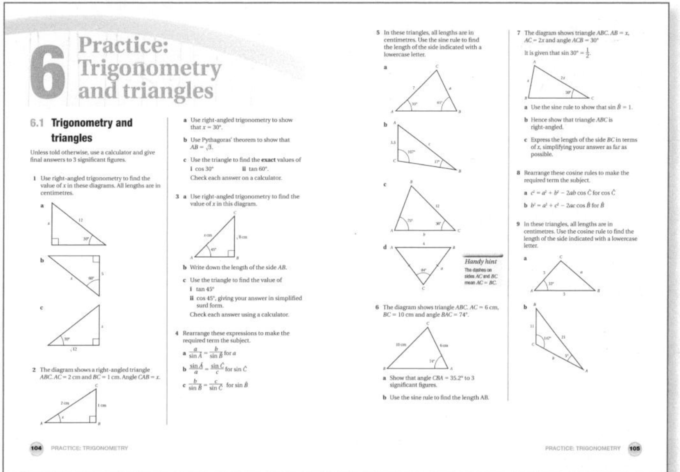
6 Practice: Trigonometry and triangles

6.1 Trigonometry and triangles

Handy hint

104 PRACTICE: TRIGONOMETRY

PRACTICE: TRIGONOMETRY 105

Key points

Look out for the blue 'Key points' boxes that highlight the most important things to remember for each topic.

Handy hints and AS Alerts

Find valuable 'Handy hint' boxes throughout the book and understand methods that are specific to AS Level with 'AS Alert!' boxes.

Common errors and Checkpoints

Avoid the common misconceptions that students regularly make with 'Common errors' boxes and discover useful ways to check your workings with 'Checkpoint' boxes.

Practice section

Increase your confidence and improve your skills with comprehensive practice sections, packed with questions, dedicated to each topic.

Exam practice

Ensure you are ready to start your AS course by taking the tear-out exam paper at the back of the book.

Answers

Find all the answers to the practice exercises at the back of the book. The answers and mark scheme for the exam paper can be found in the downloadable teacher material, available to purchase at www.collinseducation.com

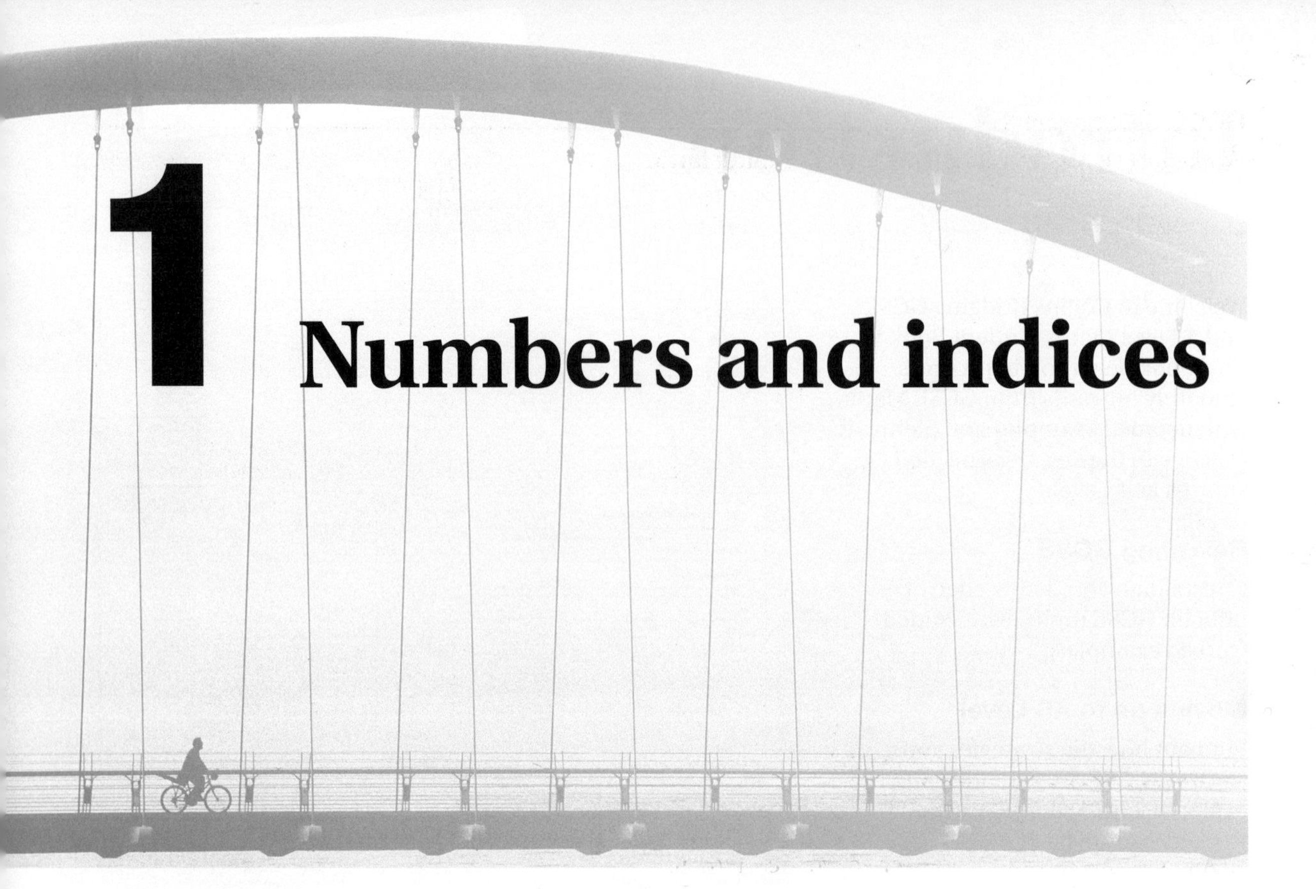

1 Numbers and indices

1.1 Fractions

What you should already know:

- how to add, subtract, multiply and divide by fractions.
- how to convert between mixed numbers and top-heavy fractions.

In this section you will learn:

- how to substitute a fraction into a formula.

Revisiting GCSE

In GCSE Maths you performed calculations with fractions. The numbers involved may have been **top-heavy** (or **improper**) fractions, such as $\frac{11}{7}$ or **mixed numbers** such as $2\frac{3}{5}$.
You should be confident in converting mixed numbers into top-heavy fractions.

For example, $2\frac{3}{5} = 2 + \frac{3}{5}$

Use a common denominator of 5: $= \frac{2 \times \mathbf{5}}{1 \times \mathbf{5}} + \frac{3}{5}$

$= \frac{10}{5} + \frac{3}{5}$

Add the numerators: $= \frac{13}{5}$

So $2\frac{3}{5} = \frac{13}{5}$

Collins Student Book

Bridging GCSE and A Level Maths

Mark Rowland

Contents

* Integration is covered in the accompanying teacher material on www.collinseducation.com

GCSE C Example 1

Work out $\frac{7}{3} \div \frac{4}{9}$. Give your answer in its simplest form.

Working

You multiply $\frac{7}{3}$ by the **reciprocal** of $\frac{4}{9}$.

$$\frac{7}{3} \div \frac{4}{9} = \frac{7}{3} \times \mathbf{\frac{9}{4}}$$

Cancel the common factor 3: $= \frac{7}{{}_1\cancel{3}} \times \frac{{}^3\cancel{9}}{4}$

Bring the numerators together:
Bring the denominators together: $= \frac{7 \times 3}{1 \times 4}$

So $\frac{7}{3} \div \frac{4}{9} = \frac{21}{4}$

Handy hint

'Reciprocate' means 'invert' or 'turn upside down'.

The reciprocal of $\frac{4}{9}$ is $\mathbf{\frac{9}{4}}$.

Make sure you only invert the fraction you are dividing by.

Handy hint

Common factors reduce the size of the numbers you are working with.

Without cancellation, this calculation is $\frac{7 \times 9}{3 \times 4} = \frac{63}{12}$.

Key point

The rules for fraction arithmetic are:

$$\frac{a}{b} \times \frac{c}{d} = \frac{ac}{bd} \qquad \frac{a}{b} \div \frac{c}{d} = \frac{a}{b} \times \frac{d}{c}$$

$$\frac{a}{b} + \frac{c}{b} = \frac{a+c}{b} \qquad \frac{a}{b} - \frac{c}{b} = \frac{a-c}{b}$$

Handy hint

You **must** make the denominators equal before you can add or subtract fractions.

In AS Maths, you will be expected to work confidently with fractions. They are often needed when simplifying, or finding the value of, an expression. You will generally find it easier to work with top-heavy fractions rather than mixed numbers.

For example, it is not easy to work out the value of $\frac{4}{1\frac{1}{2}}$ without a calculator.

However, by writing $1\frac{1}{2}$ as $\frac{3}{2}$, you can easily work out this answer.

$$\frac{4}{\left(\frac{3}{2}\right)} = \frac{4}{1} \times \mathbf{\frac{2}{3}}$$
$$= \frac{4 \times 2}{1 \times 3}$$
$$= \frac{8}{3}$$

Moving on to AS Level

AS Level Example 2

Using the formula $y = \frac{3}{4}x + \frac{5}{2}$, find the value of y when $x = 2$.

Working

You substitute $x = 2$ into the formula.

Write 2 as the top-heavy fraction $\frac{2}{1}$.

Formula: $y = \frac{3}{4}x + \frac{5}{2}$

$= \frac{3}{4} \times \frac{2}{1} + \frac{5}{2}$

Cancel the common factor 2: $= \frac{3}{{}_2\cancel{4}} \times \frac{{}^1\cancel{2}}{1} + \frac{5}{2}$

Simplify the product: $= \frac{3}{2} + \frac{5}{2}$

Add the numerators: $= \frac{3+5}{2}$

$= \frac{8}{2}$

So $y = 4$ when $x = 2$

Common error

$\frac{3}{4} \times 2$ does **not** equal $\frac{3 \times 2}{4 \times 2}$.

To avoid this type of error, write 2 as $\frac{2}{1}$.

AS Level Example 3

Using the formula $y = \frac{5x}{3} - \frac{1}{2}$, find the value of y when $x = \frac{2}{3}$.

Working

You can write $\frac{5x}{3}$ as $\frac{5}{3} \times \frac{x}{1}$, or simply as $\frac{5}{3}x$.

So the formula is $y = \frac{5}{3}x - \frac{1}{2}$.

You substitute $x = \frac{2}{3}$ into this formula.

Formula: $y = \frac{5}{3}x - \frac{1}{2}$

$= \frac{5}{3} \times \frac{2}{3} - \frac{1}{2}$

$= \frac{10}{9} - \frac{1}{2}$

Subtract the fractions by using a common denominator 18:

$= \frac{10 \times \mathbf{2}}{9 \times \mathbf{2}} - \frac{1 \times \mathbf{9}}{2 \times \mathbf{9}}$

$= \frac{20}{18} - \frac{9}{18}$

Subtract the fractions: $= \frac{11}{18}$

So $y = \frac{11}{18}$ when $x = \frac{2}{3}$

Handy hint

The coefficient of x is $\frac{5}{3}$

You can separate the coefficient from the x-term.

Key point

The expression $\frac{ax}{b}$ can be written as $\frac{a}{b}x$, where a and b are any numbers, $b \neq 0$.

Handy hint

'$b \neq 0$' means that b is not zero.

Taking it further

You will need to be confident in using fractions as you progress through your course. In particular, you use them when working with straight-line equations.

1.2 Surds

What you should already know:

- how to re-write a surd such as $\sqrt{8}$.
- how to rationalise the denominator of a fraction such as $\frac{10}{\sqrt{5}}$.

In this section you will learn:

- how to use the rules of surds to simplify more complex expressions.
- how to rationalise more complex denominators in fractions.
- how to calculate with numbers and expressions involving surds.
- the meaning of 'exact form'.

Revisiting GCSE

A surd is a number of the form $\sqrt{a}$ where a is not a square number.

For example, $\sqrt{8}$ is a surd. A decimal approximation for $\sqrt{8}$ is 2.828427125 but you can never exactly describe a surd using decimals.

However, $\sqrt{9}$ is *not* a surd because you can write down its exact value, 3.

Handy hint

You should be familiar with the square numbers: 1, 4, 9, 16, 25, 36…

Handy hint

You can assume $\sqrt{9}$ means the positive square root of 9.

You need to know the rules of surds:

Key point

The rules of surds are: $\sqrt{ab} = \sqrt{a} \times \sqrt{b}$, $\sqrt{\frac{a}{b}} = \frac{\sqrt{a}}{\sqrt{b}}$ (where $b \neq 0$).

These rules allow you to break down a surd. For example,

$$\begin{aligned}\sqrt{8} &= \sqrt{4 \times 2} \\ &= \sqrt{4} \times \sqrt{2} \\ &= 2 \times \sqrt{2}\end{aligned}$$

It is not possible to break down $\sqrt{2}$ any further using integers.
$\sqrt{8}$ has been expressed in **simplified surd form**.

GCSE A Example 4

a Express $\sqrt{12} + \sqrt{48}$ in the form $k\sqrt{3}$ where k is an integer.

b Simplify $\frac{21}{\sqrt{3}}$ by rationalising the denominator.

Common error

$\sqrt{12} + \sqrt{48}$ is NOT the same as $\sqrt{12 + 48}$ – check this on a calculator.

Working

a You need to re-write $\sqrt{12}$ using factors of 12.

$$\begin{aligned}\sqrt{12} &= \sqrt{4 \times 3} \\ &= \sqrt{4} \times \sqrt{3} \\ &= 2\sqrt{3}\end{aligned}$$

Handy hint

Use a square number which is a factor of 12 (i.e. 4).

Re-write $\sqrt{48}$ using factors of 48.

48 has lots of factors. Look for the *greatest* square number which is a factor of 48 (i.e. 16).

Handy hint

Before answering a surd question, make a list of the first few square numbers (i.e. 1, 4, 9, **16**, 25…).

$$\begin{aligned}\sqrt{48} &= \sqrt{16 \times 3} \\ &= \sqrt{16} \times \sqrt{3} \\ &= 4\sqrt{3}\end{aligned}$$

So, $\sqrt{12} + \sqrt{48} = 2\sqrt{3} + 4\sqrt{3}$

$= 6\sqrt{3}$ (so $k = 6$)

Handy hint

You can add 'like' surds.

b The denominator of $\frac{21}{\sqrt{3}}$ is the surd $\sqrt{3}$.

You need to find a fraction with the same value as $\frac{21}{\sqrt{3}}$ which does not have a surd in its denominator.

Since $\sqrt{3} \times \sqrt{3} = 3$, you multiply the numerator and denominator of this fraction by $\sqrt{3}$.

$$\begin{aligned}\frac{21}{\sqrt{3}} &= \frac{21 \times \sqrt{\mathbf{3}}}{\sqrt{3} \times \sqrt{\mathbf{3}}} \\ &= \frac{21\sqrt{\mathbf{3}}}{3} \\ &= \frac{21}{3}\sqrt{3} \\ &= 7\sqrt{3}\end{aligned}$$

Handy hint

You must multiply top **and** bottom by $\sqrt{3}$.

Key point

$\sqrt{a} \times \sqrt{a} = a$ for any number $a \geqslant 0$. Exact form is the same as giving an answer in surd form.

Moving on to AS Level

In AS Maths you will need to be able to simplify expressions involving surds. Also, you sometimes need to give answers in **exact form**, rather than using decimals.

AS Level Example 5

a Show that $(3 + \sqrt{7})(3 - \sqrt{7})$ is an integer, stating its value.

b Hence express $\frac{1 + \sqrt{7}}{3 + \sqrt{7}}$ in the form $a + b\sqrt{7}$ where a and b are integers to be stated.

AS Alert!

Part **b** does not tell you how to start.

Working

a You need to expand the brackets and then simplify the numbers.

$$(3 + \sqrt{7})(3 - \sqrt{7}) = 3 \times 3 - 3\sqrt{7} + 3\sqrt{7} - \sqrt{7} \times \sqrt{7}$$
$$= 9 - 7$$
$$= 2\text{, which is an integer}$$

Handy hint

Write $\sqrt{7} \times 3$ as $3\sqrt{7}$.

Use $\sqrt{7} \times \sqrt{7} = 7$.

b You need to rationalise the denominator of the fraction $\frac{1 + \sqrt{7}}{3 + \sqrt{7}}$

AS Alert!

Part **b** is a 'Hence' question – this means you should make use of part **a**.

This can be done by multiplying the numerator and denominator by $(3 - \sqrt{7})$.

So, $$\frac{1 + \sqrt{7}}{3 + \sqrt{7}} = \frac{(1 + \sqrt{7}) \times (\mathbf{3} - \sqrt{\mathbf{7}})}{(3 + \sqrt{7}) \times (\mathbf{3} - \sqrt{\mathbf{7}})}$$

$$= \frac{3 - \sqrt{7} + 3\sqrt{7} - 7}{2}$$

Handy hint

You should use the result of part **a** to simplify the denominator.

Combine numbers and like surds: $$= \frac{-4 + 2\sqrt{7}}{2}$$

$$= \frac{-4}{2} + \frac{2\sqrt{7}}{2}$$

$$= -2 + \sqrt{7}$$

Compare this answer to the form $a + b\sqrt{7}$ given in the question: $-2 + \sqrt{7}$ looks like $a + b\sqrt{7}$ where $a = -2$, $b = 1$.

AS Alert!

You must write down the value of a and the value of b, as directed by the question.

Key point

If a and b are integers, then $(a + \sqrt{b})(a - \sqrt{b})$ is also an integer, and equals $a^2 - b$.

To rationalise the denominator of a fraction of the form $\frac{N}{a + \sqrt{b}}$, multiply top and bottom by $(a - \sqrt{b})$.

To rationalise the denominator of a fraction of the form $\frac{N}{a - \sqrt{b}}$, multiply top and bottom by $(a + \sqrt{b})$.

Taking it further

Surds appear in many areas of AS Maths, such as solving quadratic equations and calculating distances between points.

1.3 Indices

What you should already know:

- how to simplify expressions such as $(a^2b)^2$.
- how to calculate values such as 3^{-2} or $4^{\frac{1}{2}}$.

In this section you will learn:

- how to work out the value of more complicated expressions using the rules of indices.

In GCSE Maths you would have used various rules of indices to simplify an expression. For example, in the expression $a^2 \times a^3$ you can add the indices together so that $a^2 \times a^3$ simplifies to a^5.

This works because $a^2 \times a^3 = (a \times a) \times (a \times a \times a)$

$= a^5$

Key point

Here are some results involving indices:

1. $a^m \times a^n = a^{m+n}$
2. $\frac{a^m}{a^n} = a^{m-n}$ for a not zero
3. $(a^m)^n = (a^n)^m = a^{mn}$
4. $a^n \times b^n = (ab)^n$
5. $\frac{a^n}{b^n} = \left(\frac{a}{b}\right)^n$ for b not zero

Handy hint

Rule **3** means $(a^m)^n$ and $(a^n)^m$ are equal to each other and are also equal to a^{mn}.

GCSE B Example 6

Simplify $(2p^3q^2)^2$.

Working

You can write rule **4** as $(ab)^n = a^n \times b^n$

So $(2p^3q^2)^2 = 2^2 \times (p^3)^2 \times (q^2)^2$

Use rule 3 on each bracket: $= 4 \times (p^{3 \times 2}) \times (q^{2 \times 2})$

Simplify the indices: $= 4p^6q^4$

So $(2p^3q^2)^2$ simplifies to $4p^6q^4$.

Handy hint

Rule **4** can be extended to three or more terms.

You will also have met the **definition** of negative and fractional indices.

For example, 2^{-3} means $\frac{1}{2^3}$ and so 2^{-3} has value $\frac{1}{8}$.

Similarly, $4^{\frac{1}{2}}$ means the positive square root of 4, and so $4^{\frac{1}{2}}$ has value 2.

$3^0 = 1$ and, in general, $a^0 = 1$ for any number a.

GCSE A Example 7

Find the value of $3^{-2} \times 8^{\frac{1}{3}}$.

Working

You apply the definitions to each expression.

$$3^{-2} = \frac{1}{3^2} = \frac{1}{9} \qquad 8^{\frac{1}{3}} = \sqrt[3]{8} = 2$$

So $3^{-2} \times 8^{\frac{1}{3}} = \frac{1}{9} \times 2 = \frac{2}{9}$

Handy hint

$8^{\frac{1}{3}} = (2^3)^{\frac{1}{3}}$
$= 2^{3 \times \frac{1}{3}}$
$= 2$

So $8^{\frac{1}{3}} = \sqrt[3]{8}$, the cube root of 8.

Moving on to AS Level

Key point

Here are some general definitions which you need to know for AS Maths.

6 $a^{-n} = \frac{1}{a^n}$ where n is any integer.

7 $a^{\frac{1}{n}} = \sqrt[n]{a}$ where n is any positive integer.

In AS Maths you will be expected to evaluate expressions involving indices, possibly without the use of a calculator.

AS Level Example 8

a Find the value of $16^{-\frac{3}{2}}$.

b Simplify $\frac{12x^2}{2\sqrt{x}}$.

c Express $\frac{4x^3 - 3}{x}$ in the form $ax^2 - bx^n$, stating the value of the constants a, b and n.

AS Alert!

You can see that the numbers and expressions are more complex at this level. You need to combine several rules of indices to answer these questions.

Working

a Using rule **6** you can write $16^{-\frac{3}{2}}$ as $\frac{1}{16^{\frac{3}{2}}}$.

Now concentrate on the denominator.

You can write $16^{\frac{3}{2}}$ as $\left(16^{\frac{1}{2}}\right)^3$ by using rule **3**.

Now use rule **7**: $\left(16^{\frac{1}{2}}\right)^3 = (\sqrt{16})^3 = (4)^3 = 64$

So $16^{-\frac{3}{2}} = \frac{1}{64}$

Handy hint

Deal with the negative sign on the index first.

Handy hint

Using rule **3**, you could also work out $16^{\frac{3}{2}}$ as $(16^3)^{\frac{1}{2}}$, but this would mean having to find 16^3.

b You can write $\frac{12x^2}{2\sqrt{x}}$ as $\frac{12}{2} \times \frac{x^2}{\sqrt{x}}$ which simplifies to $6\frac{x^2}{\sqrt{x}}$.

Now express the denominator as a power of x.

So $6\frac{x^2}{\sqrt{x}} = 6\frac{x^2}{x^{\frac{1}{2}}}$

Use rule **2**: $= 6x^{2-\frac{1}{2}}$

$= 6x^{\frac{3}{2}}$

So $\dfrac{12x^2}{2\sqrt{x}}$ simplifies to $6x^{\frac{3}{2}}$.

Handy hint

Although $2 - \frac{1}{2} = 1\frac{1}{2}$, you should avoid writing an index using a mixed number.

c You must split the fraction up into two parts:

$$\frac{4x^3 - 3}{x} = \frac{4x^3}{x} - \frac{3}{x}$$

Separate the numbers from the x-terms: $= 4 \times \dfrac{x^3}{x} - 3 \times \dfrac{1}{x}$

Use rules of indices: $= 4x^{3-1} - 3x^{-1}$

$= 4x^2 - 3x^{-1}$

So $\dfrac{4x^3 - 3}{x} = 4x^2 - 3x^{-1}$

Compare $4x^2 - 3x^{-1}$ with $ax^2 - bx^n$: $a = 4$, $b = 3$, $n = -1$

Common error

b is **not** -3. Comparing terms, $-b = -3$ so $b = 3$.

Key point

$a^{\frac{m}{n}} = (\sqrt[n]{a})^m$

$= \sqrt[n]{a^m}$

When calculating $a^{\frac{m}{n}}$ it is usually easier to use the result $(\sqrt[n]{a})^m$ rather than $\sqrt[n]{a^m}$

Taking it further

Indices appear in many AS Maths topics, but in particular you will need to be confident in using them when studying differentiation and integration.

2.1 Basic algebra

What you should already know:

- how to simplify expressions such as $a(a + 4) - (a + 1)^2$ or $(2a^3b)^2$.
- how to rearrange a formula such as $A = \pi r^2$ to make r the subject.

In this section you will learn:

- how to deal with more complicated expressions and formulae.

Revisiting GCSE

In GCSE Maths you practised expanding brackets, factorising and simplifying expressions.

GCSE B Example 1

a Factorise fully $12a^5b + 6a^2b^3$.

b Rearrange $A = \dfrac{h(a + b)}{2}$ to make b the subject of the formula.

Working

a You look for common factors of numbers and terms.

In the expression $12a^5b + 6a^2b^3$

the numbers are 12 and 6: the highest *common* factor is 6 because $12 = 6 \times 2$.

the powers of a are a^5 and a^2: the highest *common* power is a^2 because $a^5 = a^2 \times a^3$.

the powers of b are b and b^3: the highest *common* power is b^1 because $b^3 = b^1 \times b^2$.

So the highest common factor of the expression is $6a^2b$.

In fully factorised form, $12a^5b + 6a^2b^3 = 6a^2b(2a^3 + b^2)$.

Handy hint

The smaller of the two powers gives the common factor.

Checkpoint

Multiplying out the brackets gives you the expression you started with.

b Formula to rearrange: $A = \frac{h(a+b)}{2}$

Multiply both sides by 2: $2A = h(a+b)$

Divide both sides by h: $\frac{2A}{h} = a + b$

Subtract a from both sides: $\frac{2A}{h} - a = b$

So $b = \frac{2A}{h} - a$

> *Handy hint*
>
> You could multiply out the bracket $h(a+b)$ but this makes rearranging for b slightly harder.

Moving on to AS Level

In AS Maths, as well as being confident in expanding and factorising expressions, you will also need to be able to express an equation in a particular form for use in another part of a question. There is usually more than one way of doing this.

AS Level Example 2

Express $\frac{8x^5 + 6x^2}{2x^2}$ in the form $Ax^3 + B$, stating the value of the constants A and B.

Working

In one method you fully factorise the numerator.

$$8x^5 + 6x^2 = 2x^2(4x^3 + 3)$$

So
$$\frac{8x^5 + 6x^2}{2x^2} = \frac{2x^2(4x^3 + 3)}{2x^2}$$
$$= 4x^3 + 3$$

Compare $4x^3 + 3$ with $Ax^3 + B$: $A = 4$, $B = 3$.

Alternatively, you could have divided each term in the numerator by $2x^2$.

$$\frac{8x^5 + 6x^2}{2x^2} = \frac{8x^5}{2x^2} + \frac{6x^2}{2x^2}$$
$$= 4x^3 + 3$$

> *Handy hint*
>
> You can cancel the $2x^2$ terms:
> $\frac{2x^2(4x^3 + 3)}{2x^2} = \frac{2x^2}{2x^2} \times (4x^3 + 3)$
> $= 4x^3 + 3$

> *AS Alert!*
>
> State the value of A and the value of B, as directed in the question.

> *Handy hint*
>
> Use rules of indices
> e.g. $\frac{8x^5}{2x^2} = \frac{8}{2}x^{5-2} = 4x^3$.

You may also be required to find a formula and then rearrange it, where the required letter appears on *both* sides of the formula.

AS Level Example 3

The diagram shows a right-angled triangle ABC, where $BC = p$ and $AC = p + 1$.

If $s = \sin \hat{A}$ find an expression for p in terms of s.

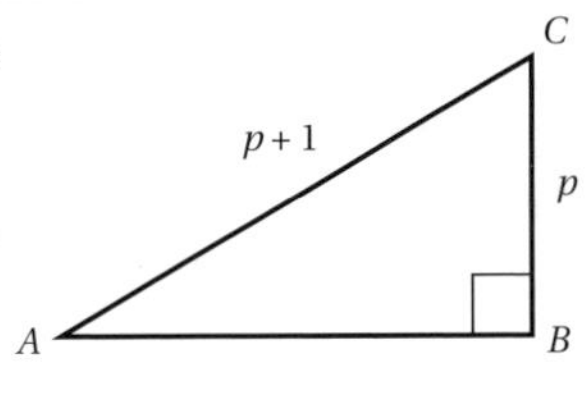

> *AS Alert!*
>
> 'An expression for p' means find a formula whose subject is p.

Working

You need to use right-angled trigonometry.

$$\sin \hat{A} = \frac{\text{opposite}}{\text{hypotenuse}} \text{ so } s = \frac{p}{p+1}$$

So the formula $s = \frac{p}{p+1}$ must be rearranged to make p the subject.

Formula to rearrange: $s = \frac{p}{p+1}$

Multiply both sides by $(p + 1)$: $s(p+1) = p$

Expand brackets: $sp + s = p$

Gather terms in p to one side and terms in only s to the other:

$$sp - p = -s$$

Factorise the left-hand side: $(s-1)p = -s$

> *Common error*
>
> $\frac{p}{p+1}$ is **not** equal to $\frac{p}{p} + \frac{1}{p}$.

> *Common error*
>
> A rearrangement of $s = \frac{p}{p+1}$ is **not** $sp + 1 = p$.
>
> You must use a bracket around the term $p + 1$.

Divide both sides by $(s - 1)$: $p = \frac{-s}{s - 1}$

So an expression for p in terms of s is $p = \frac{s}{1 - s}$.

Handy hint

Tidy up the answer by multiplying top and bottom of $\frac{-s}{s-1}$ by (-1).

Taking it further

Being able to manipulate algebraic expressions quickly is an important skill in AS Maths. Most questions depend on these techniques!

2.2 Solving linear equations

What you should already know:

- how to solve an equation such as $2(3a + 1) = a + 7$.
- how to solve a pair of simultaneous equations such as $2x + 3y = 11$ and $3x - 2y = 10$.

In this section you will learn:

- how to solve more complicated equations.
- how to solve an equation as part of a more complex problem.

Revisiting GCSE

In GCSE Maths you learnt how to solve an equation by applying reverse operations to each side of the equation.

GCSE C Example 4

Solve the equation $\frac{2x}{3} + 1 = 7$.

Working

You need to 'undo' the equation by applying reverse operations to each side, so that adding becomes subtracting, dividing becomes multiplying etc.

Equation to solve: $\frac{2x}{3} + 1 = 7$

Subtract 1 from both sides: $\frac{2x}{3} = 6$

Multiply both sides by 3: $2x = 18$

Divide both sides by 2: $x = 9$

The solution is $x = 9$.

Checkpoint

Substitute $x = 9$ into the equation

$\frac{2x}{3} + 1 = \frac{2(9)}{3} + 1$

$= \frac{18}{3} + 1 = 6 + 1$

$= 7$ (checked).

Moving on to AS Level

In AS Maths, you often need to solve an equation as part of a more complex problem. This means you have to be able to apply the rules with confidence. The equation may involve an unknown in the numerator and denominator of a fraction.

AS Level Example 5

Solve the equation $\frac{3t}{(t + 3)} = 2$.

Common error

$\frac{3t}{t + 3}$ is **not** the same as $\frac{3t}{t} + \frac{3t}{3}$.

Working

You need to multiply both sides by the denominator $(t + 3)$.

Equation to solve: $\frac{3t}{(t + 3)} = 2$

Multiply both sides by $(t + 3)$: $3t = 2(t + 3)$

Expand the bracket: $3t = 2t + 6$

Subtract $2t$ from both sides: $t = 6$

So the solution is $t = 6$.

> *Common error*
>
> The rearrangement is **not** $3t = 2t + 3$. Put brackets around the denominator to avoid this error.

Here is a more complex problem which involves several steps and is more typical of an AS-style question.

AS Level Example 6

The diagram shows a rectangle $ABCD$.

$AB = x$, $BC = 2x + 2$, where all dimensions are in centimetres.

The perimeter of this rectangle is 34 cm.

a Show that the side AB has length 5 cm.

b Hence find the length of the diagonal AC.

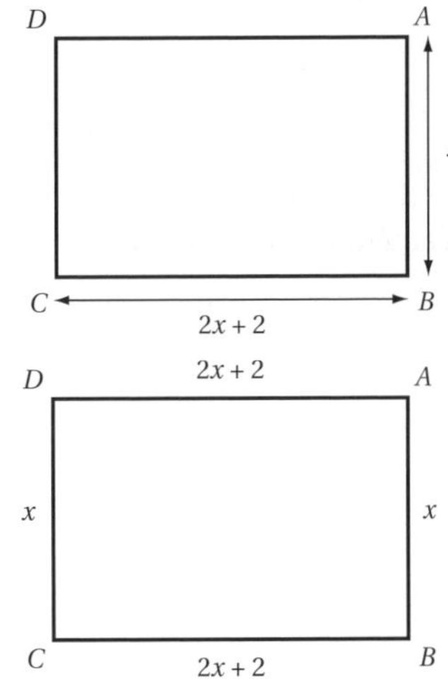

> *AS Alert!*
>
> Many AS questions use the phrase 'Show that'. Here, this means you must find the value of x, even though you know it will work out to be 5!

Working

a You must form and then solve an equation involving x.

Because $ABCD$ is a rectangle, its perimeter in terms of x is

$$AB + BC + CD + DA = x + (2x + 2) + x + (2x + 2)$$
$$= 6x + 4$$

Since the perimeter is 34 cm, the equation to solve is

$$6x + 4 = 34$$

Subtract 4 from both sides: $6x = 30$

Divide both sides by 6: $x = 5$

So the length of the side AB is 5 cm, as required.

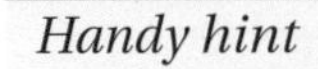

> *Handy hint*
>
> You must not use the given answer $x = 5$ anywhere in your working for part **a**.

b You can use the value $x = 5$ to find the length of side BC.

$$BC = 2x + 2$$
$$= 2(5) + 2$$
$$= 12 \text{ cm}$$

> *Handy hint*
>
> You are allowed to use $x = 5$ in part **b**. This means you can attempt this part of the question even if you could not do part **a**.

Now use Pythagoras' theorem to find the length of the diagonal AC.

$$AC^2 = AB^2 + BC^2$$
$$= 5^2 + 12^2$$
$$= 25 + 144$$
$$= 169$$

So $AC = \sqrt{169}$
$$= 13$$

The diagonal AC has length 13 cm.

> *AS Alert!*
>
> You were not told to find the length of BC. In AS Maths you must learn to think for yourself!

> *Handy hint*
>
> C1 is a non-calculator paper. You need to know your squares and square roots!

Taking it further

Equation-solving features in almost every aspect of AS Maths. You may need to be able to form your own equations and solve them without being told exactly what to do at each step.

2.3 Linear inequalities

What you should already know:

- how to solve an equation such as $3x + 2 = 17$.
- how to solve a simple inequality such as $2x + 1 \leqslant 7$.

In this section you will learn:

- how to solve more complicated inequalities.

Revisiting GCSE

Inequalities are expressions showing a range of possible values. For example, you have to be at least 17 years old before you can obtain a driving licence. So if a person has a driving licence then you can write $A \geqslant 17$, where A is the person's age.

GCSE B Example 7

Solve the inequality $6x + 3 < 2x + 15$.

Working

You apply the same rules as you would when solving an equation.

Inequality to solve: $6x + 3 < 2x + 15$

Subtract $2x$ from both sides: $4x + 3 < 15$

Subtract 3 from both sides: $4x < 12$

Divide both sides by 4: $x < 3$

So the solution is $x < 3$

Handy hint

Do **not** replace the inequality symbol with an equals sign – you may forget to change it back again at the end!

Handy hint

$x < 3$ means x is less than 3. It also means 3 is greater than x.

Moving on to AS Level

In AS Maths questions, inequalities often involve fractions and negative coefficients.

AS Level Example 8

Find the set of values of x for which $\frac{2 - 3x}{4} \leqslant 2$.

Working

You can start by clearing the fraction.

Inequality to solve: $\frac{2 - 3x}{4} \leqslant 2$

Multiply both sides by 4: $2 - 3x \leqslant 8$

Subtract 2 from both sides: $-3x \leqslant 6$

Divide both sides by (-3) **and reverse the inequality sign**:

$$x \geqslant \frac{6}{-3} = -2$$

So the set of values for which $\frac{2 - 3x}{4} \leqslant 2$ is $x \geqslant -2$

AS Alert!

Inequality questions often use the phrase 'Find the set of values for which…'. This simply means 'solve the inequality…'.

Checkpoint

Test your solution by choosing a value of $x \geqslant -2$, e.g. $x = 0$. When $x = 0$, $\frac{2 - 3x}{4} = \frac{2 - 3(0)}{4} = \frac{2}{4} < 2$ (checked).

Key points

1. When you multiply or divide each side of an inequality by a negative number, you must reverse the inequality symbol.
2. In AS Maths, you will meet expressions involving two inequalities. For example, the expression $2 \leqslant x \leqslant 4$ means x can take any value between 2 and 4 (inclusive). Some possible values of x for which $2 \leqslant x \leqslant 4$ are 2, 2.5, 3.05, π, and 4.
3. $a \leqslant x \leqslant b$ means x can take any value between a and b (inclusive).

   $a < x < b$ means x can take any value between a and b except a and b themselves.

AS Level Example 9

a Solve the inequality $3 \leqslant 2x + 5 < 13$.

b Hence write down the set of values of x for which $-2 < x \leqslant 3$ **and** $.3 \leqslant 2x + 5 < 13$.

> *AS Alert!*
> You may be required to solve 'simultaneous' inequalities.

> *AS Alert!*
> Part **b** is a 'Hence' question – this means you should make use of part **a**.

Working

a You can treat both inequalities simultaneously.

Inequality to solve: $3 \leqslant 2x + 5 < 13$

Subtract 5 from *all* sides: $3 - \mathbf{5} \leqslant 2x + 5 - \mathbf{5} < 13 - \mathbf{5}$

So $-2 \leqslant 2x < 8$

Divide *all* sides by 2: $-1 \leqslant x < 4$

So the solution is $-1 \leqslant x < 4$

> *Handy hint*
> $-1 \leqslant x < 4$ means x lies between -1 and 4. x can equal -1 but cannot equal 4.

b You can represent each inequality on a number line.

The closed circle ● means you include an end value, the open circle ○ means you exclude an end value.

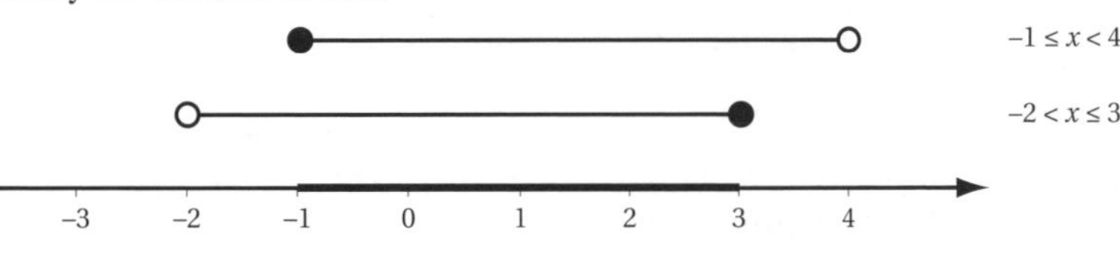

In the diagram, the top line represents the solution to $3 \leqslant 2x + 5 < 13$ as found in part **a**.

The lines overlap when x is between -1 and 3 inclusive.

So the set of values which satisfy both inequalities is $-1 \leqslant x \leqslant 3$.

> *Handy hint*
> Use appropriate mathematical notation and clearly state your answer. Do not leave it 'hidden' on the diagram.

Taking it further

You will need to be able to solve inequalities such as $x^2 - x < 6$. You also need to solve inequalities to find out when a quadratic equation has real solutions. Solving inequalities also helps find the set of values of x for which a graph is increasing.

2.4 Forming expressions

What you should already know:

- how to form an expression from a diagram.

In this section you will learn:

- how to work with two variables on a diagram.

Revisiting GCSE

In GCSE Maths you worked with formulae for perimeters, areas and volumes of various shapes. For example, the volume V of a cylinder with base radius r and height h is given by the formula $V = \pi r^2 h$.

GCSE C Example 10

The diagram shows a rectangular allotment and path.

The width of the path is x m and the rectangle $ABCD$ has dimensions 6 m × 11 m.

Find an expression for the perimeter of the allotment.

Give your answer in factorised form.

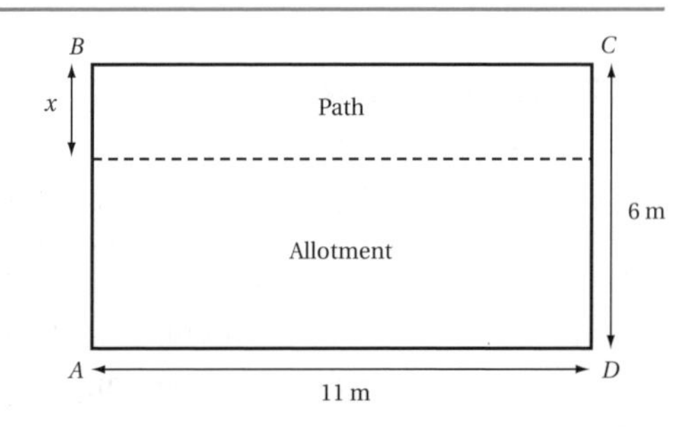

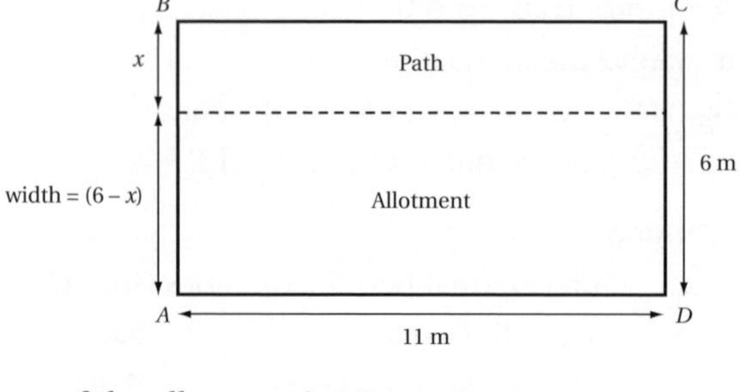

Working

You need to describe the width of the allotment in terms of x.

The allotment has width $(6 - x)$ m.

The perimeter of the allotment

$= 11 + (6 - x) + 11 + (6 - x)$

Gather like terms: $= 11 + 6 + 11 + 6 - x - x$

Simplify: $= 34 - 2x$

So, in factorised form, an expression for the perimeter of the allotment is $2(17 - x)$ m.

You can turn this expression into an algebraic equation by writing $P = 2(17 - x)$, where P is the perimeter of the allotment.

Moving on to AS Level

In AS Maths, you may have to deal with a shape which has more than one unknown dimension.

AS Level Example 11

The diagram shows a rectangle $PQRS$ where $PQ = x$, $QR = y$.
All lengths are in centimetres.

The rectangle has perimeter 12 cm and area A cm^2.

Show that $A = 6x - x^2$.

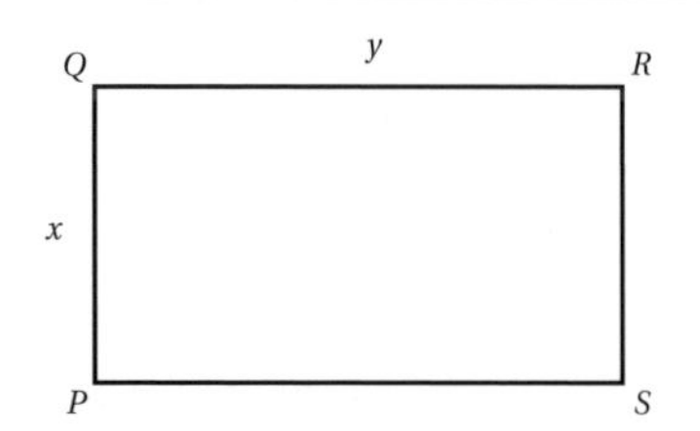

Working

You know that the area of the rectangle is $A = xy$, and you are trying to show that $A = 6x - x^2$.

So you need to substitute y with an expression in terms of x.

To find this expression, you must use the given information that the rectangle has perimeter 12 cm.

Perimeter $= x + y + x + y$

$= 2x + 2y$

So you know that $2x + 2y = 12$

Rearrange this equation to make y the subject: $y = \dfrac{12 - 2x}{2}$

Now substitute y with this expression in $A = xy$.

So $A = xy$

$= x\left(\dfrac{12 - 2x}{2}\right)$

Divide out by 2: $= x\left(\dfrac{12}{2} - \dfrac{2x}{2}\right)$

$= x(6 - x)$

Expand the bracket: $= 6x - x^2$

Hence $A = 6x - x^2$, as required.

Handy hint

You must use all the information given, in this case, perimeter = 12 cm.

Common error

Do not try to solve the equation $2x + 2y = 12$. (You do not have enough information to solve it!)

Taking it further

Being able to form expressions appears in practical problems when trying to maximise or minimise quantities such as surface areas or volumes.

3 Coordinate geometry 1

3.1 Straight-line graphs

What you should already know:

- how to draw the graph of a straight line such as $y = 2x + 5$ by plotting points.
- how to identify the gradient and y-intercept of a line using its equation and/or graph.

In this section you will learn:

- how to sketch the graph of a straight line.
- how to find the gradient and y-intercept from the equation of a line.
- how to express a line equation in different ways.

Revisiting GCSE

In GCSE Maths, you may have drawn a straight-line graph by plotting some points onto a grid. This gives an accurate picture of the graph.

GCSE C Example 1

a Draw the line $y = 2x + 5$ for $-3 \leqslant x \leqslant 2$.

b Write down the value of x where this line crosses the x-axis.

Handy hint

$-3 \leqslant x \leqslant 2$ means x takes *any* value between -3 and 2 (inclusive) but in this question you are only using integer values.

Working

a You can make a table of values using the integers x for which $-3 \leqslant x \leqslant 2$.

x	-3	-2	-1	0	1	2
$y = 2x + 5$	-1	1	3	5	7	9

Plot these points on a grid and join them to make a straight line (see next page).

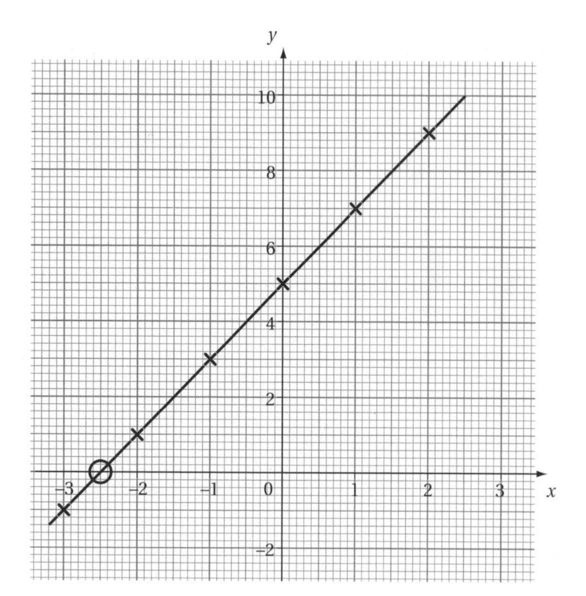

b By inspecting the graph you can see the line crosses the x-axis when $x = -2.5$.

The line drawn in part **a** has equation $y = 2x + 5$.

The constant term is 5. This is the *y-intercept* of the line and tells you where the line crosses the y-axis.

The coefficient of x is 2. This is the *gradient* of the line and tells you how steep the line is. A gradient of 2 means y increases by 2 whenever x increases by 1 (i.e. 1 along, 2 up).

Handy hint

A gradient is by how much y changes when x increases by 1 unit.

Checkpoint

Check these two features for the line drawn in part **a**.

Key point

Every straight line has an equation $y = mx + c$, where m is the gradient of the line, and c is the y-intercept of the line.

Moving on to AS Level

In AS Maths, the equation of a line may appear in different forms. You need to be confident in converting between forms.

AS Level Example 2

The line L_1 has equation $y = \frac{5}{2} - \frac{3}{4}x$.

a Express the equation of L_1 in the form $ay + bx = c$ where a, b and c are integers to be stated.

The line L_2 has equation $2y - 5x = 11$.

b Find the gradient of the line L_2.

AS Alert!

C1 questions often ask for answers in the form shown in part **a**.

AS Alert!

Subscripts are useful for distinguishing between the two lines. L_1 is the first line, L_2 is the second line.

Working

a You need to rearrange $y = \frac{5}{2} - \frac{3}{4}x$ into the form $ay + bx = c$.

Equation to rearrange: $y = \frac{5}{2} - \frac{3}{4}x$

Multiply all terms by 4: $4y = 10 - 3x$

Add $3x$ to both sides: $4y + 3x = 10$

Compare $4y + 3x = 10$ with $ay + bx = c$: $a = 4$, $b = 3$, $c = 10$ are integers.

Handy hint

Multiply through by lowest common multiple of 2 and 4 to remove all fractions.

This makes no difference to the points on the line.

b You need to rearrange $2y - 5x = 11$ into the form $y = mx + c$.

Equation to rearrange: $2y - 5x = 11$

Add $5x$ to both sides: $2y = 11 + 5x$

Divide all terms by 2: $y = \frac{11}{2} + \frac{5}{2}x$

Compare $y = \frac{11}{2} + \frac{5}{2}x$ with $y = mx + c$: $m = \frac{5}{2}$.

The gradient of L_2 is $\frac{5}{2}$.

Handy hint

You can only read off the gradient and y-intercept of a line when its equation is in the form $y = mx + c$.

Common error

The gradient of the line $y = \frac{11}{2} + \frac{5}{2}x$ is **not** $\frac{11}{2}$.

In this form, the coefficient of x gives the gradient.

In AS Maths you should not spend time plotting points on graph paper when solving problems involving straight lines – you must learn how to **sketch** a line.

A sketch: does not use scales on the axes
does not involve plotting points
only needs to display the main features of a graph.

Handy hint

A sketch is useful for checking that answers look reasonable.

Figure 1 shows a sketch of the line with equation $y = 4x - 3$. This line has a positive gradient and a negative y-intercept. One possible equation for the line in Figure 2 is $y = 6 - 2x$, since the line has a positive y-intercept and a negative gradient.

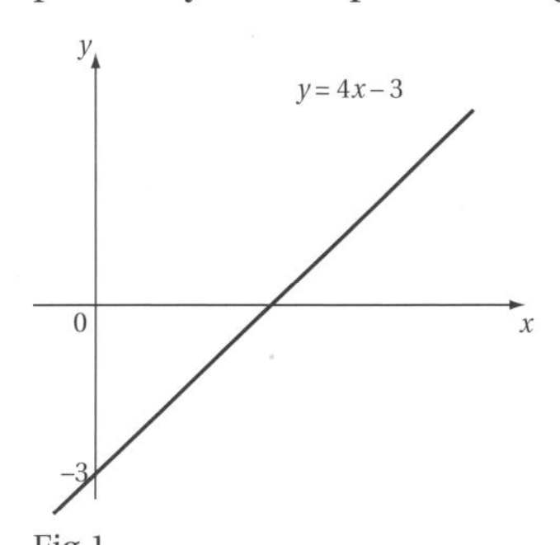

Fig.1

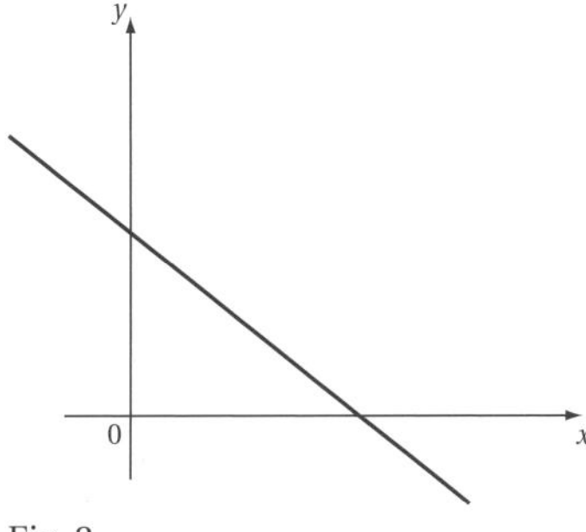

Fig. 2

In AS Maths, questions on straight lines often involve a geometrical problem.

AS Level Example 3

Find the area of the triangle formed by the x-axis, the y-axis and the line with equation $y = 7 - 3x$.

Working

You need to sketch the line so that you can visualise the triangle.

The line $y = 7 - 3x$ has gradient -3 and so slopes 'downwards'.

The line has y-intercept 7.

AS Alert!

You need to be able to translate the words into a diagram.

The diagram shows a sketch of this line. The points A and B are where the line crosses each axis, and O is the origin.

You need to find the dimensions of triangle OAB.

The line crosses the x-axis at the point where $y = 0$.

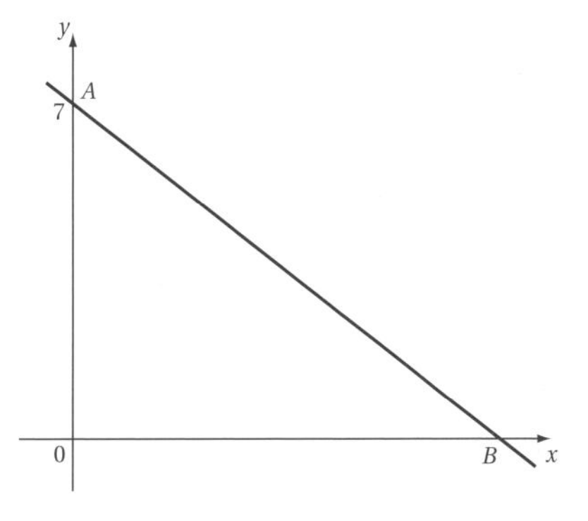

AS Alert!

The question did not ask you to find the coordinates of point B – you need to figure this out for yourself!

Equation to solve: $0 = 7 - 3x$

Add $3x$ to both sides: $3x = 7$

Divide both sides by 3: $x = \frac{7}{3}$

So the triangle OAB has height $OA = 7$ and base $OB = \frac{7}{3}$.

The area of this triangle is $= \frac{1}{2} \times OB \times OA$

$$= \frac{1}{2} \times \frac{7}{3} \times \frac{7}{1} = \frac{49}{6} \text{ units}^2$$

AS Alert!

You are expected to work with fractions. Do not convert $\frac{7}{3}$ to a decimal as it could lead to inaccuracies.

Taking it further

You will deal with straight lines extensively in the Coordinate Geometry sections of AS Maths.

3.2 The equation of a line

What you should already know:

- how to calculate the gradient of a line using two points.
- how to locate the y-intercept of a line.
- how to find the equation of a line by using its gradient and y-intercept.

In this section you will learn:

- how to find the equation of any line using two points.

Revisiting GCSE

In GCSE Maths, you learnt how to find the equation of a line by first calculating its gradient. The y-intercept was either given to you, or could easily be read off from a diagram.

GCSE C Example 4

Find the equation of the line L shown in the diagram (Figure 1).

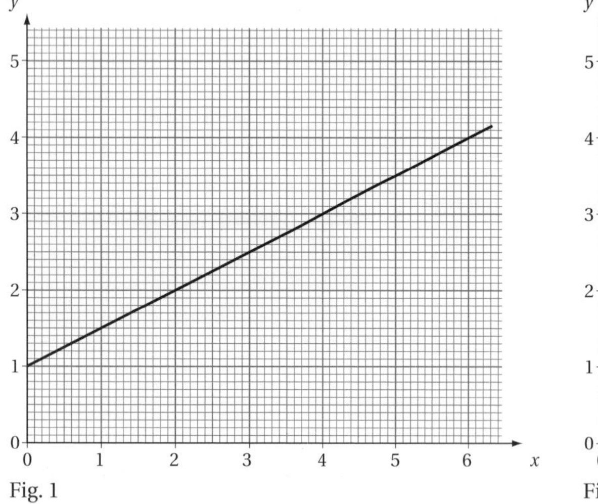

Fig. 1

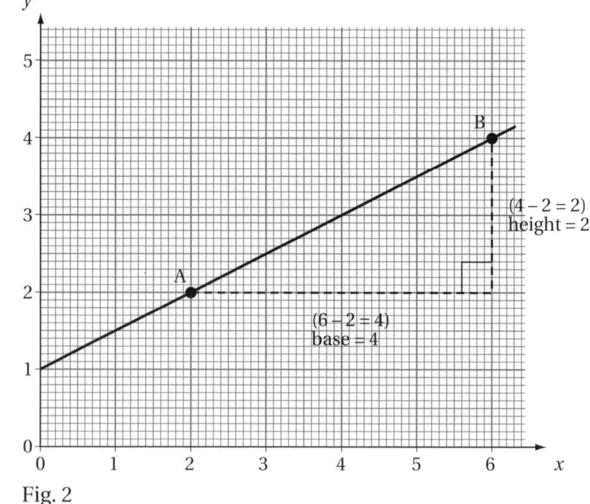

Fig. 2

Working

You know that the line has equation $y = mx + c$ where m is the gradient and c the y-intercept.

You can choose any two points on this line, say A and B (Figure 2, above), and by drawing a right-angled triangle under the line AB, you can find the gradient m.

So the gradient $m = \frac{\text{height}}{\text{base}} = \frac{4-2}{6-2}$

$= \frac{2}{4}$

$= \frac{1}{2}$

You can see from the graph that the line crosses the y-axis where $y = 1$.

So the y-intercept c is 1.

Hence the equation of the line is $y = \frac{1}{2}x + 1$.

Checkpoint

Use the coordinates $B(6,4)$.

When $x = 6$, $y = \frac{1}{2}x + 1$

$= \frac{1}{2}(6) + 1$

$= 3 + 1$

$= 4$ (checked).

Key point

A line which passes through the points $A(x_1,y_1)$ and $B(x_2,y_2)$ has gradient $m = \frac{y_2 - y_1}{x_2 - x_1}$.

Handy hint

You can think of a gradient as 'the change in y divided by the change in x'

Moving on to AS Level

The equation of the line in Example 4 was relatively easy to find because the y-intercept was given on the graph. In AS Maths, you may not be given the graph of a line and so you will have to find the y-intercept by using algebra.

AS Level Example 5

The line L passes through the points $A(1,9)$ and $B(3,17)$.

a Find the equation of L.

b Show that the point $D(-2,-3)$ lies on L.

AS Alert!

In this question you are not given the y-intercept. You will have to work it out!

Working

a You can find the gradient m of L by using the formula

$m = \frac{y_2 - y_1}{x_2 - x_1}$ where $(x_1,y_1) = (1,9)$ and $(x_2,y_2) = (3,17)$.

$\frac{y_2 - y_1}{x_2 - x_1} = \frac{17 - 9}{3 - 1}$

$= \frac{8}{2}$

So $m = 4$ and the equation of L is $y = 4x + c$.

To find c, you rearrange this equation to make c the subject.

Equation: $y = 4x + c$

Subtract $4x$ from both sides: $y - 4x = c$

The equation $y - 4x = c$ is satisfied by any point on L.

So substitute the coordinates of $A(1,9)$ into this equation to find c.

Use $A(1,9)$: $x = 1$, $y = 9$: $c = y - 4x$

$= 9 - 4(1)$

$= 5$

So the equation of L is $y = 4x + 5$.

Handy hint

You would get the same answer for m if instead you made $(x_2,y_2) = (1,9)$ and $(x_1,y_1) = (3,17)$.

Checkpoint

You would get the same answer for c if you used point $B(3,17)$ instead. Try this for yourself.

b You need to substitute the x-coordinate of point D into the equation for L.

The point $D(-2,-3)$ has x-coordinate -2.

Equation: $y = 4x + 5$

Substitute x with -2: $= 4(-2) + 5$

$= -8 + 5$

$= -3$

This answer equals the y-coordinate of D, so this means $D(-2,-3)$ must lie on L.

Key point

If a line has gradient m and passes through the point $A(x_1,y_1)$ then the y-intercept of the line is given by $c = y_1 - mx_1$.

Taking it further

You need to be able to find the equation of a straight line and use the line to solve problems in Coordinate Geometry. Lines also feature in the study of differentiation.

3.3 Mid-points and distances

What you should already know:

- how to find the coordinates of the mid-point of the line joining two points.
- how to find the distance between two points.

In this section you will learn:

- how to use a formula to find the mid-point of a line joining points A and B.
- how to use a formula to find the distance between two points.
- how to apply these formulae to problems.

Revisiting GCSE

In GCSE Maths, you may have found the mid-point of the line joining the points A and B by using common sense or reading off values from a graph.

GCSE A Example 6

The grid shows a line passing through the points $A(1,5)$ and $B(7,1)$.

M is the mid-point of AB.

a Find the coordinates of M.

b Find the distance AM. Leave your answer as a surd.

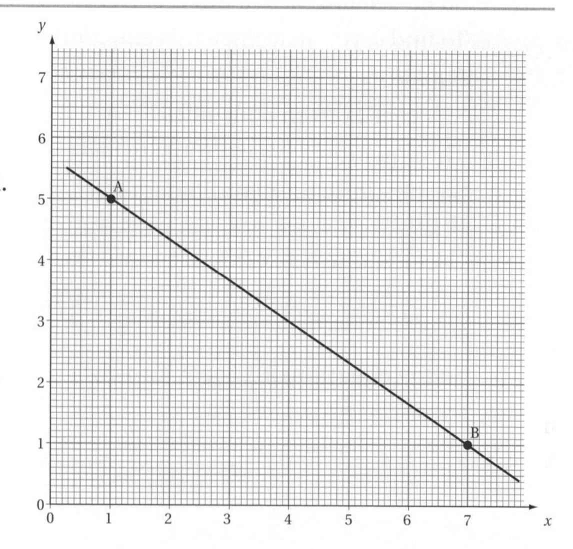

Working

a The x-coordinates of A and B are 1 and 7. So the x-coordinate of M is 4 because 4 is half-way between 1 and 7.

Similarly, looking at the y-coordinates of these points, half-way between 1 and 5 is 3 so the y-coordinate of M must be 3.

So M has coordinates (4,3).

b Draw a right-angled triangle under the line from A to M.

Using Pythagoras' theorem,

$AM^2 = 2^2 + 3^2$

$= 13$

So the distance $AM = \sqrt{13}$.

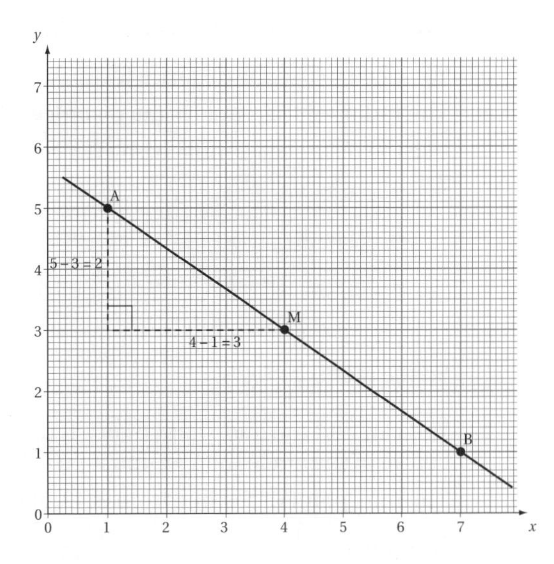

Moving on to AS Level

In AS Maths you need to be able to use formulae for finding mid-points and distances.

Key point

The mid-point M of the points $A(x_1,y_1)$ and $B(x_2,y_2)$ has coordinates $\left(\frac{x_1 + x_2}{2}, \frac{y_1 + y_2}{2}\right)$.

The length of the line from A to B is given by $AB = \sqrt{(x_2 - x_1)^2 + (y_2 - y_1)^2}$.

Checkpoint

Check that these formulae work for the points A and B in Example 6.

The diagram explains why these results are true.

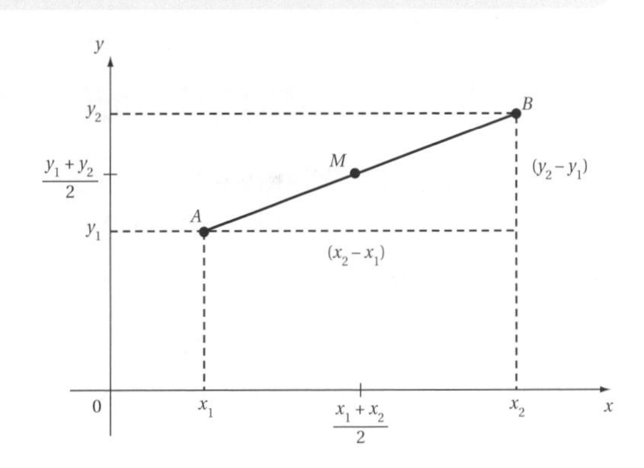

AS Level Example 7

The diagram shows a circle with centre $A(6,4)$. Point $B(11,5)$ lies on this circle.

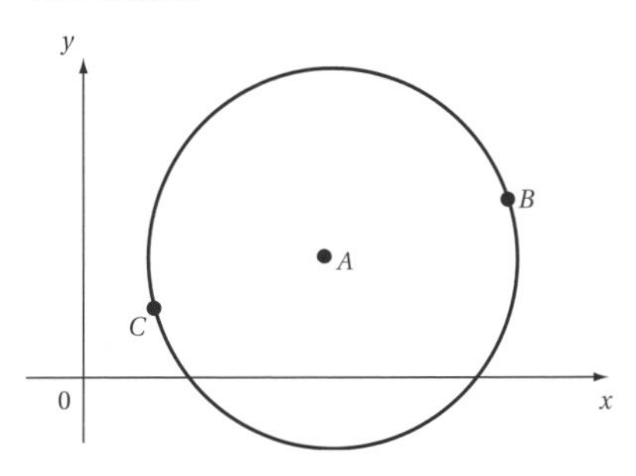

a Find the radius r of this circle.

b Verify that BC is a diameter of this circle, where C has coordinates (1,3).

AS Alert!

Circles can be centred anywhere, not just at the origin.

AS Alert!

'Verify' means you can use the coordinates of C in your working. This is different from a 'Show that' question – refer to Example 6 in section 2.2.

Working

a You need to use the distance formula.

The radius of this circle is the distance AB.

Use the distance formula with $(x_1,y_1) = (6,4)$ and $(x_2,y_2) = (11,5)$.

$$AB = \sqrt{(x_2 - x_1)^2 + (y_2 - y_1)^2}$$
$$= \sqrt{(11 - 6)^2 + (5 - 4)^2}$$
$$= \sqrt{(5)^2 + (1)^2}$$
$$= \sqrt{25 + 1}$$

So the radius $r = \sqrt{26}$.

> *Common error*
>
> $\sqrt{5^2 + 1^2}$ is **not** equal to 5 + 1.

b You can find the mid-point of BC. If this mid-point is the centre of the circle then BC must be a diameter.

The mid-point of the line from $B(11,5)$ to $C(1,3)$ has

coordinates $\left(\frac{11 + 1}{2}, \frac{5 + 3}{2}\right) = \left(\frac{12}{2}, \frac{8}{2}\right)$

$= (6,4)$

which are the coordinates of the centre $A(6,4)$.

This means BC must be a diameter of the circle, as required.

> *Handy hint*
>
> In this context, a diameter is any line joining two points on the circle and which passes through the centre.

> *Handy hint*
>
> You can also verify BC is a diameter by showing the distance $BC = 2r = 2\sqrt{26}$.

Taking it further

You will use the formulae for mid-points and distances when answering Coordinate Geometry problems in C1. They are also used in Circle Geometry questions.

3.4 Parallel and perpendicular lines

What you should already know:

- how to tell when two lines are parallel to each other or perpendicular to each other.
- how to find the gradient of a line which is perpendicular to a given line.

In this section you will learn:

- how to apply these results to more complex problems.

Revisiting GCSE

In GCSE Maths, questions on parallel or perpendicular lines are often broken down into several parts. You should be familiar with the two main results about parallel or perpendicular lines.

Key point

Two lines are parallel if they have equal gradients.

Two lines are perpendicular if they are at right-angles.

If a line L has gradient m then:

- the gradient of any line parallel to L is also m
- the gradient of any line perpendicular to L is $-\frac{1}{m}$, provided $m \neq 0$.

GCSE A Example 8

The grid shows the line $y = 2x + 1$ and a line L which passes through point $A(0,3)$.

These two lines are parallel.

a Write down the equation of line L.

b Write down the gradient of a line perpendicular to L.

c Write down the equation of the line perpendicular to L passing through $A(0,3)$.

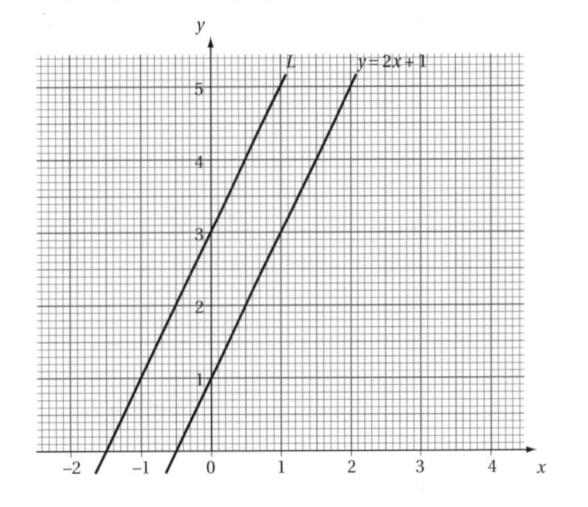

Working

a You know that the equation of L is $y = mx + c$ where m is the gradient and c the y-intercept.

The gradient of the line $y = 2x + 1$ is 2.

As the two lines are parallel, this means their gradients are equal and so $m = 2$.

You can clearly read off the y-intercept 3 of L from the graph.

So $c = 3$.

Hence the equation of L is $y = 2x + 3$.

b You know the gradient of L is 2.

So any line perpendicular to L has gradient $-\frac{1}{2}$.

c The line perpendicular to L and which passes through the point $A(0,3)$ has gradient $-\frac{1}{2}$ and y-intercept 3.

So the equation of this line is $y = -\frac{1}{2}x + 3$ (see dotted line on the diagram).

> ***Handy hint***
>
> Express m as a fraction:
>
> $m = \frac{2}{1} \xrightarrow[\text{and change sign}]{\text{invert}} = -\frac{1}{2}$.

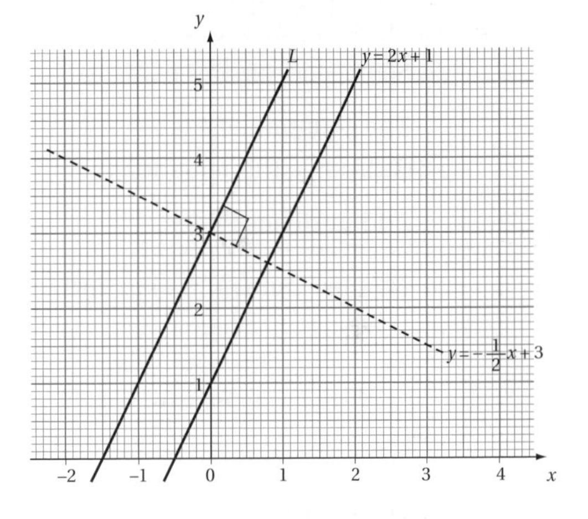

Moving on to AS Level

In AS Maths, questions on this topic are not usually broken down in this way. Also, the question may not even contain a diagram and so a sketch may be needed.

AS Level Example 9

The points A and B have coordinates (3,4) and (9,8) respectively.

a Find the gradient of AB.

b Hence find the equation of the perpendicular bisector of the line AB.

Give your answer in the form $ay + bx = c$ for integers a, b and c.

> ***AS Alert!***
>
> The word 'respectively' means you pair items off in the obvious order. So here, this means A has coordinates (3,4) and B has coordinates (9,8).

Working

a You need to use the formula $m = \frac{y_2 - y_1}{x_2 - x_1}$ where $(x_1,y_1) = (3,4)$ and $(x_2,y_2) = (9,8)$

So $m = \frac{8 - 4}{9 - 3}$

$= \frac{4}{6}$

The gradient of the line AB is $\frac{2}{3}$.

> ***Handy hint***
>
> Read the question carefully – it asks only for the gradient of AB, not the equation of AB.

b You should sketch a graph to help visualise the problem.

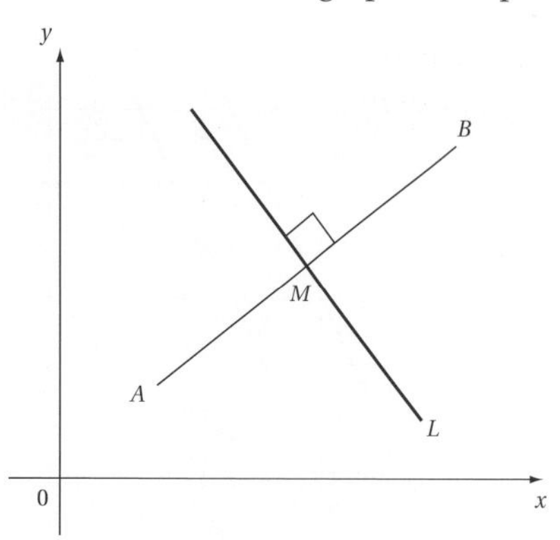

On the sketch, point M is the mid-point of AB and the line L is the perpendicular bisector of AB.

To find the equation of L you need to find

i the gradient of L and **ii** the coordinates of M.

i The gradient of AB is $\frac{2}{3}$.

L is perpendicular to AB so the gradient of L is $-\frac{3}{2}$.

ii The mid-point of $A(3,4)$ and $B(9,8)$ has coordinates $\left(\frac{3+9}{2}, \frac{4+8}{2}\right)$.

So the coordinates of M are (6,6).

The equation of L is $y = -\frac{3}{2}x + c$.

Rearrange this equation to make c the subject.

Add $\frac{3}{2}x$ to both sides: $y + \frac{3}{2}x = c$

Now use point $M(6,6)$ to find c:

$$\begin{aligned} c &= y + \frac{3}{2}x \\ &= 6 + \frac{3}{2}(6) \\ &= 15 \end{aligned}$$

So the equation of L is $y = -\frac{3}{2}x + 15$

Finally, multiply this equation through by 2 to produce integers: $2y = -3x + 30$.

So the equation of the perpendicular bisector of AB is $2y + 3x = 30$ where $a = 2$, $b = 3$ and $c = 30$ are integers.

Handy hint

The perpendicular bisector of AB is the line perpendicular to AB and which passes through the mid-point of AB.

Handy hint

Although it is only a sketch, it is important to correctly position the points $A(3,4)$ and $B(9,8)$ relative to the origin O and to each other.

AS Alert!

You need to develop strategies for solving problems in AS Maths. Try to ask yourself general questions such as 'what do I need to know in order to find the equation of a line'?

Handy hint

$\frac{2}{3} \xrightarrow[\text{and change sign}]{\text{invert}} = -\frac{3}{2}$.

Handy hint

Remember to multiply *all* terms by 2, including the 15.

Handy hint

Make sure you rearrange the answer into the required form. Here, this means gathering x and y terms to the same side.

Taking it further

Parallel and perpendicular lines feature heavily in Coordinate Geometry and differentiation problems.

3.5 Intersections of lines

What you should already know:

- how to read off from a graph the coordinates where two lines intersect.

In this section you will learn:

- how to use algebra to find where two lines intersect.

Revisiting GCSE

At GCSE level, you may have used a graphical method to find where two lines intersect. This method usually only gives an estimate for the coordinates of the intersection point.

GCSE C Example 10

The diagram (Figure 1) shows a drawing of the line $y = 4x + 1$ and the line $y = 14.5 - 2x$. These lines intersect at point P.

Write down the coordinates of P.

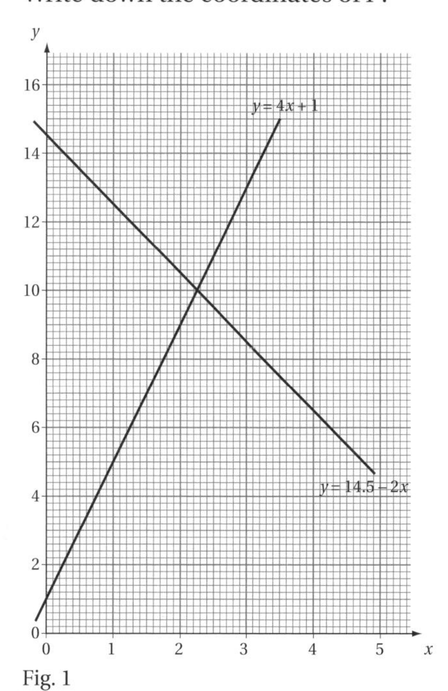

Fig. 1

Fig. 2

Working

By reading off from the diagram, you can see the lines appear to intersect when $x = 2.25$ and $y = 10$ (Figure 2).

So the coordinates of P appear to be (2.25,10).

> *Handy hint*
>
> No matter how carefully you look, you cannot be sure that you have correctly read off the coordinates of P. For example, the actual coordinates of P could be (2.24,9.98).

Moving on to AS Level

In AS Maths questions on this topic, you may not be able to read off coordinates of intersection points accurately. Instead you will be expected to use algebra to find **exactly** where two lines intersect. One method is to use **substitution**.

Example 11 uses the same pair of lines as shown in Example 10.

AS Level Example 11

Use algebra to find the coordinates of the point P where the lines $y = 4x + 1$ and $y = 14.5 - 2x$ intersect.

Working

The equations are $\quad y = 4x + 1 \quad (1)$

and $\quad y = 14.5 - 2x \quad (2)$

To find where these lines intersect, you can substitute the letter y in equation (2) with the expression $(4x + 1)$ from equation (1).

So $4x + 1 = 14.5 - 2x$

Add $2x$ to both sides: $6x + 1 = 14.5$

Subtract 1 from both sides: $6x = 13.5$

Divide both sides by 6: $x = \frac{13.5}{6} = 2.25$

Now substitute $x = 2.25$ into $y = 4x + 1$ to find the y-coordinate of P.

$y = 4(2.25) + 1$

$= 10$

So the coordinates of P are (2.25,10).

Handy hint

$x = 2.25$ means the x-coordinate of P is 2.25.

Handy hint

By using algebra, you can now be sure that the coordinates of P are (2.25,10) – see Example 10.

Key point

Use the method of substitution to find where two lines intersect.

AS questions on this topic tend to include several parts and may require you to find the area of a shape such as a triangle drawn in the x–y plane.

AS Level Example 12

The diagram shows the line $y = 7 - x$ and the line $4y + 3x = 24$.

The lines intersect at point A. The line $y = 7 - x$ crosses the y-axis at point B.

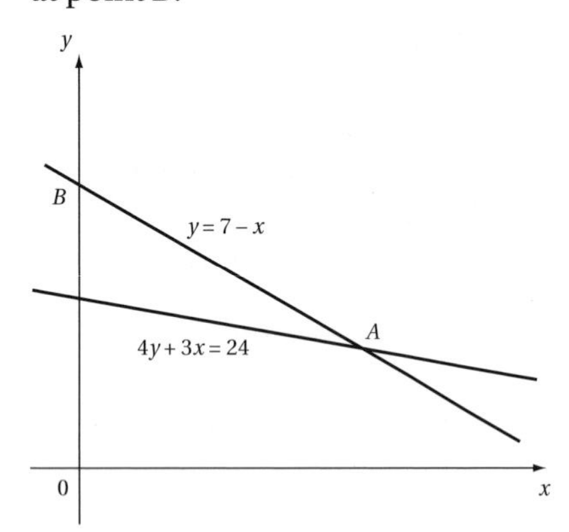

a Find the coordinates of point A.

b Write down the coordinates of point B.

c Hence find the area of triangle OAB, where O is the origin.

AS Alert!

In part **b**, 'write down' suggests the answer should be obvious – look out for this phrase in any exam or test.

Working

a The equations are $y = 7 - x$ (1)

$4y + 3x = 24$ (2)

To find the x-coordinate of A, you substitute y in equation (2) with the expression $(7 - x)$ in equation (1).

Equation to solve: $4(7 - x) + 3x = 24$

Expand the brackets: $28 - 4x + 3x = 24$

Simplify: $28 - x = 24$

Solve: $x = 4$

So the x-coordinate of A is 4.

Handy hint

You could also find the x-coordinate of A by solving the simultaneous equations

$y + x = 7$ (1)

$4y + 3x = 24$ (2)

Handy hint

You must use brackets when substituting for y.

Now substitute $x = 4$ into the line $y = 7 - x$ to find the y-coordinate of A.

$$\begin{aligned} y &= 7 - x \\ &= 7 - 4 \\ &= 3 \end{aligned}$$

So A has coordinates (4,3).

b The line $y = 7 - x$ has y-intercept 7.
So B has coordinates (0,7).

c You need to find the base and height of triangle OAB.
Put the x-coordinate of A onto the sketch.

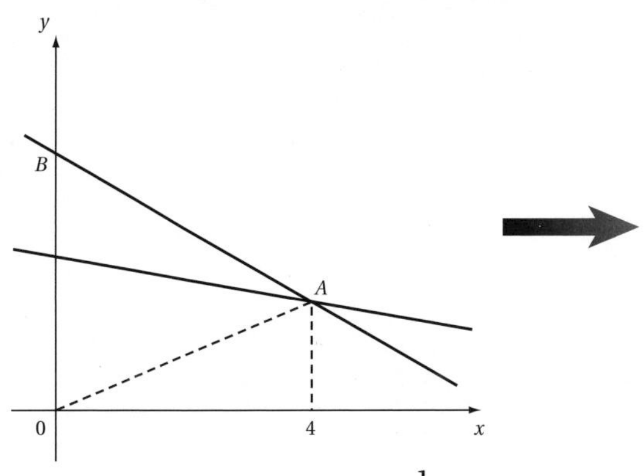

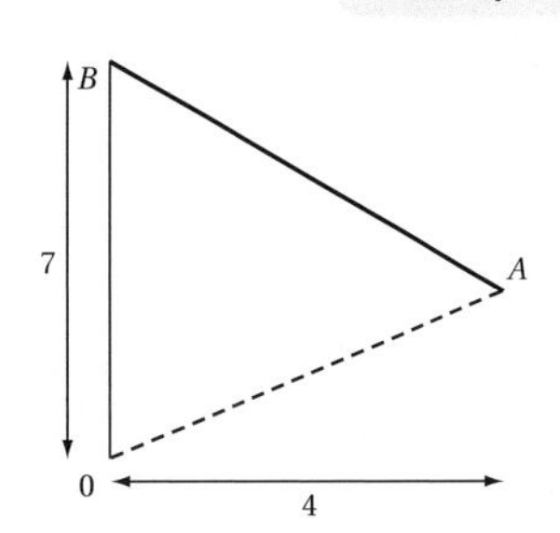

The area of triangle OAB is $= \frac{1}{2}$ base × height

$= \frac{1}{2}(7)(4)$

So the area of this triangle is 14.

Handy hint

Use whichever equation is easiest for finding y, in this case, the equation $y = 7 - x$.

Checkpoint

Use equation (2) to check the answer $A(4,3)$:

$$\begin{aligned} 4y + 3x &= 4(3) + 3(4) \\ &= 12 + 12 = 24 \text{ (checked)} \end{aligned}$$

Handy hint

In part **b**, make sure you write down the coordinates of B – do not just write down the y-intercept 7.

Handy hint

Think of the base as the line OB, which has length 7.

Turn the page 90° anticlockwise if it helps!

Taking it further

In AS Maths, you will need to be able to find the intersection points of a curve and a line, or even a pair of curves. The method of substitution can be used to do this.

4 Algebra 2

4.1 Solving a quadratic equation by factorising

What you should already know:

- how to factorise a quadratic expression such as $x^2 - 5x + 6$.
- how to solve a quadratic equation such as $2x^2 + 9x + 4 = 0$ by factorising.

In this section you will learn:

- how to form and solve a quadratic equation as part of a more complex question.

Revisiting GCSE

In GCSE Maths, you saw that factorising helps you to find the solutions of a quadratic equation. Often the questions are broken down into steps and you are told to use factorisation.

GCSE B Example 1

a Factorise $x^2 + 4x - 21$.

b Hence solve the equation $x^2 + 4x - 21 = 0$.

Working

a Expression to factorise: $x^2 + 4x - 21$

Look at the constant 21 (ignore the negative sign).

Can you find two integers whose product is 21? (yes: try 3 and 7)

Look at the middle term $4x$.

Can you make 4 by using 3 and 7? (yes: $4 = 7 - 3$)

So $x^2 + 4x - 21 = (x + 7)(x - 3)$.

Handy hint

Always focus on the constant term first.

b The equation $x^2 + 4x - 21 = 0$ can be written as $(x + 7)(x - 3) = 0$.

So either $x + 7 = 0$ or $x - 3 = 0$

The solutions are $x = -7$ or $x = 3$

Handy hint

Make each bracket equal to zero and then solve each equation.

Checkpoint

Check that $x = 3$ satisfies the equation $x^2 + 4x - 21 = 0$

$(3)^2 + 4(3) - 21 = 9 + 12 - 21$
$= 21 - 21$
$= 0$ (checked).

Key point

The equation $(x - p)(x - q) = 0$ has solutions $x = p, x = q$.

This method only works if one side of the equation is zero.

Moving on to AS Level

In AS Maths, you often need to form and solve a quadratic equation in order to proceed with the question. You will not be told how to solve the equation.

AS Level Example 2

The diagram shows a rectangle *ABCD*, where $AB = 2x$ and $BC = 2x + 7$. All lengths are in metres.

The area of this rectangle is 30 m^2.

a Show that $2x^2 + 7x - 15 = 0$.

b Find the perimeter of this rectangle.

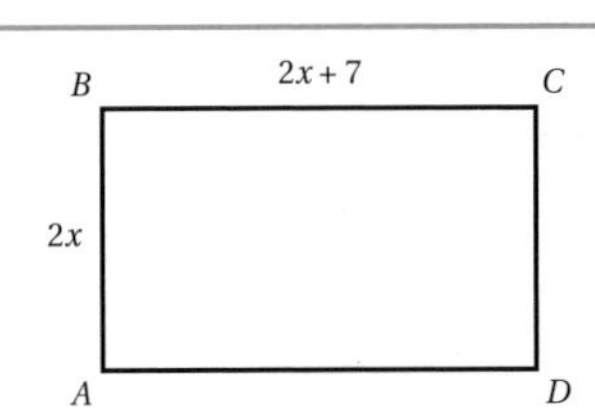

Working

a You need to find an expression for the area of this rectangle.

The area of *ABCD* = length × width

$= 2x(2x + 7)$

Now use the given information.

The area is 30 m^2 so: $2x(2x + 7) = 30$

Divide both sides by 2: $x(2x + 7) = 15$

Expand the bracket: $2x^2 + 7x = 15$

Subtract 15 from both sides: $2x^2 + 7x - 15 = 0$

So $2x^2 + 7x - 15 = 0$, as required.

b To find the perimeter, you need to first find x.

Equation to solve: $2x^2 + 7x - 15 = 0$

This factorises as $(2x - 3)(x + 5) = 0$

So either $2x - 3 = 0$ or $x + 5 = 0$

Hence $x = \frac{3}{2}$ or $x = -5$ (ignore, as $x > 0$).

The rectangle has width $2x = 2\left(\frac{3}{2}\right)$

$= 3$ metres

and length $2x + 7 = 2\left(\frac{3}{2}\right) + 7$

$= 3 + 7 = 10$ metres

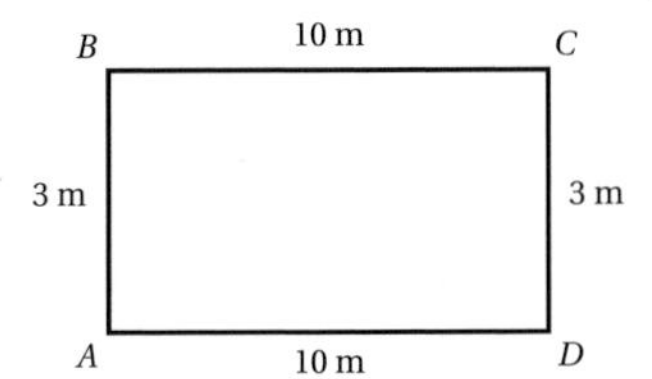

So the perimeter of *ABCD* is $3 + 10 + 3 + 10 = 26$ metres.

AS Alert!

You need to develop strategies for these types of questions. Ask yourself questions such as 'what do I need to find first in order to find the perimeter?'

Checkpoint

Expand $(2x - 3)(x + 5)$ to make sure the factorisation is correct.

Common error

A solution to $(2x - 3)(x + 5) = 0$ is **not** $x = 3$. You must set each bracket to zero to find the values of x.

AS Alert!

You may need to ignore some solutions to an equation. Although $x = -5$ is a solution, it must be ignored since x is a length and so must be positive.

Taking it further

Quadratic equations run through the whole of C1 and C2. Often they arise from working on a more complex problem and so it is important that you are able to factorise quickly and accurately.

4.2 Completing the square

What you should already know:

- how to use completing the square to solve a quadratic equation such as $x^2 - 4x + 2 = 0$.

In this section you will learn:

- how to use a completed square form to show an equation has no real solutions.
- how to find the minimum or maximum value of a quadratic expression.

Revisiting GCSE

In GCSE Maths, you may have used completing the square to solve a quadratic equation, giving your answer in surd form.

GCSE A* Example 3

Solve the equation $x^2 + 6x + 7 = 0$.

Give your answers in surd form.

Working

You can complete the square on the expression $x^2 + 6x + 7$.

Separate the constant 7: $x^2 + 6x \quad + 7$

Complete the square on $x^2 + 6x$: $(x + 3)^2 - 9 \quad + 7$

Combine the constant terms: $(x + 3)^2 - 2$

Hence $x^2 + 6x + 7 = (x + 3)^2 - 2$

So the equation $x^2 + 6x + 7 = 0$

can be written as $(x + 3)^2 - 2 = 0$

Add 2 to both sides: $(x + 3)^2 = 2$

Square root both sides: $x + 3 = \pm\sqrt{2}$

Subtract 3 from both sides: $x = \pm\sqrt{2} - 3$

So the solutions are $x = \sqrt{2} - 3$ or $x = -\sqrt{2} - 3$

Handy hint

Since the answer involves surds, do not try to use factorisation.

Checkpoint

You can check your answer by expanding the brackets.

$(x + 3)^2 - 2 = (x^2 + 6x + 9) - 2$
$= x^2 + 6x + 7$
(checked).

Handy hint

Do not expand the brackets to solve this equation – if you do, you will just go round in circles!

Handy hint

There are always two answers when you square root a positive number.

Key point

The completed square form for $x^2 + bx + c$ is $\left(x + \frac{b}{2}\right)^2 - \left(\frac{b}{2}\right)^2 + c$ where b and c are constants.

Moving on to AS Level

Every equation you solved in GCSE Maths had a solution. In AS Maths, you will meet equations which have no real solutions, which means the solutions do not lie on the real number line. Completing the square can be used to identify these equations.

AS Level Example 4

$f(x) = x^2 - 4x + 10$, where x is any real number.

a Find the minimum value of this function.

b Show that the equation $f(x) = 2x$ has no real solutions.

Working

a You need to express $x^2 - 4x + 10$ in completed square form.

$$x^2 - 4x + 10 = (x - 2)^2 - 4 + 10$$
$$= (x - 2)^2 + 6$$

Any number when squared is greater or equal to 0.

So $(x - 2)^2 \geqslant 0$ for all values of x.

Add 6 to both sides: $(x - 2)^2 + 6 \geqslant 6$ for all values of x.

So $f(x) \geqslant 6$ which means the minimum value of $f(x)$ is 6.

b You need to turn the expression $f(x) = 2x$ into a quadratic equation.

$f(x) = 2x$ means $x^2 - 4x + 10 = 2x$.

Subtract $2x$ from both sides: $x^2 - 6x + 10 = 0$

Express in completed square form $(x - 3)^2 - 9 + 10 = 0$

So $(x - 3)^2 + 1 = 0$

Subtract 1 from both sides: $(x - 3)^2 = -1$.

Square root both sides: $x - 3 = \pm\sqrt{-1}$.

The square root of -1 does not have a real value.

So the equation $f(x) = 2x$ does not have real solutions.

AS Alert!

Function notation is used frequently.
For example, $f(2) = (2)^2 - 4(2) + 10$
$= 4 - 8 + 10$
$= 6$

Common error

$x^2 - 4x + 10$ is **not** equal to $(x + 2)^2 - 4 + 10$

The sign on x should match the sign in the bracket.

Handy hint

In general, you must get one side equal to zero when trying to solve any quadratic equation.

Checkpoint

You can see that $(x - 3)^2 + 1$ can never equal zero because it has a minimum value of 1.

Checkpoint

Use a scientific calculator to try and find $\sqrt{-1}$. You should get an error. The number $\sqrt{-1}$ does not lie on the real number line.

Key point

For p and q constants:

- the minimum value of $(x - p)^2 + q$ is q.
- when $q > 0$, the equation $(x - p)^2 + q = 0$ does not have real solutions.

Taking it further

Completing the square is used extensively in AS Maths. It can be used to find the maximum or minimum points on a quadratic curve, and to locate the centre of a circle given its equation.

4.3 Solving a quadratic equation using the formula

What you should already know:

- how to use the quadratic formula to solve a quadratic equation such as $2x^2 - 3x - 1 = 0$.

In this section you will learn:

- how to use the quadratic formula with unknown coefficients.

Revisiting GCSE

In GCSE Maths, the quadratic formula is used to solve an equation when it is difficult to complete the square. Answers usually involve surds or are given to a certain level of accuracy.

Key point

The quadratic equation $ax^2 + bx + c = 0$ has solutions $x = \dfrac{-b \pm \sqrt{b^2 - 4ac}}{2a}$, where a, b and c are constants.

Handy hint

The formula must only be applied to an equation of the form $ax^2 + bx + c = 0$ (i.e. where one side is zero).

GCSE A Example 5

Solve the equation $2x^2 + 4x + 1 = 0$. Give your answers to 2 decimal places.

Working

Apply the formula to the equation $2x^2 + 4x + 1 = 0$, where $a = 2$, $b = 4$ and $c = 1$.

$$\text{So} \quad x = \frac{-b \pm \sqrt{b^2 - 4ac}}{2a}$$

$$= \frac{-(4) \pm \sqrt{(4)^2 - 4(2)(1)}}{2(2)}$$

$$= \frac{-4 \pm \sqrt{16 - 8}}{4}$$

$$= \frac{-4 \pm \sqrt{8}}{4}$$

$$\text{So} \quad x = \frac{-4 + \sqrt{8}}{4} \quad \text{or} \quad x = \frac{-4 - \sqrt{8}}{4}$$

$$= -0.29289\ldots \qquad = -1.70710\ldots$$

To 2 decimal places, the solutions are $x = -0.29$ and $x = -1.71$.

Common error

The answer is **not** $-4 \pm \frac{\sqrt{8}}{4}$

All terms in the numerator must be divided by 4.

Checkpoint

Check using the (ANS) on your calculator

2(ANS)^2 + 4(ANS) + 1

The answer should be 0.

Moving on to AS Level

Since C1 is a non-calculator unit, solutions found using the quadratic formula are usually required in simplified surd form. Sometimes, the formula is used with unknown coefficients rather than numbers, to determine when a real solution exists.

AS Level Example 6

$f(x) = ax^2 - 4x - 1$, where a is a non-zero constant.

a Find the possible values of a for which the equation $f(x) = 0$ has no real solutions.

b Find the solutions of the equation $f(x) = 0$ when $a = 3$. Give your answer in surd form, simplifying your answer as far as possible.

AS Alert!

Some equations you meet in AS Maths cannot be solved using real numbers.

Working

a You need to express $f(x) = 0$ as a quadratic equation.

$f(x) = 0$ means $ax^2 - 4x - 1 = 0$.

You can apply the formula to the equation $ax^2 - 4x - 1 = 0$, where $b = -4$ and $c = -1$, even though a does not have a specific value.

$$\text{So} \quad x = \frac{-b \pm \sqrt{b^2 - 4ac}}{2a}$$

$$= \frac{-(-4) \pm \sqrt{(-4)^2 - 4a(-1)}}{2a}$$

$$= \frac{4 \pm \sqrt{16 + 4a}}{2a}$$

Common error

$(-4)^2$ is **not** -16. Use brackets to avoid this type of error. Also, note that $-(-4) = 4$, and that $-4a(-1) = 4a$.

You need to focus on the term inside the square root symbol.

Since square rooting a negative number does not give a real answer, the equation $ax^2 - 4x - 1 = 0$ has no real roots when $16 + 4a < 0$.

Subtract 16 from both sides: $4a < -16$

Divide both sides by 4: $a < -4$

So no real solution to the equation $f(x) = 0$ exists when $a < -4$.

b When $a = 3$, $f(x) = 0$ means $3x^2 - 4x - 1 = 0$.

You should use the expression in part **a** with $a = 3$.

$$= \frac{4 \pm \sqrt{16 + 4a}}{6}$$

$$= \frac{4 \pm \sqrt{16 + 4(3)}}{6}$$

$$= \frac{4 \pm \sqrt{28}}{6}$$

$$= \frac{4 \pm 2\sqrt{7}}{6}$$

So the solutions are $x = \frac{2 \pm \sqrt{7}}{3}$.

Handy hint

Make use of earlier work – there is no point in re-inventing the wheel!

Handy hint

Divide *all* the integers by 2 to simplify the answers. This does not change the values of x.

Key point

- $\sqrt{d}$ is not a real number if $d < 0$.
- The equation $ax^2 + bx + c = 0$ has no real solutions if $b^2 - 4ac < 0$. The number $b^2 - 4ac$ is called the **discriminant** of the equation.

Taking it further

The quadratic formula is used frequently to solve quadratic equations that do not easily factorise and for which completing the square would be awkward. For example, $3x^2 - 11x + 2 = 0$.

You will also use the formula to investigate equations which have no real solutions.

4.4 Solving linear and non-linear equations simultaneously

What you should already know:

- how to solve a pair of equations where one is linear and one is non-linear, such as $y = x + 3$ and $x^2 + y^2 = 5$.

In this section you will learn:

- how to solve more complicated types of simultaneous equations.

Revisiting GCSE

An equation is linear if it only involves a term in x, a term in y and a constant, such as $y = 2x + 3$. A non-linear equation may involve, for example, a term in x^2 or y^2. So, $x^2 + y^2 = 5$ is an example of a *non*-linear equation.

In GCSE Maths, you may have used the method of substitution to solve equations simultaneously.

Handy hint

See Section 3.5 for the method of substitution.

GCSE A* Example 7

The diagram shows the circle $x^2 + y^2 = 10$ and the line $y = x + 2$.

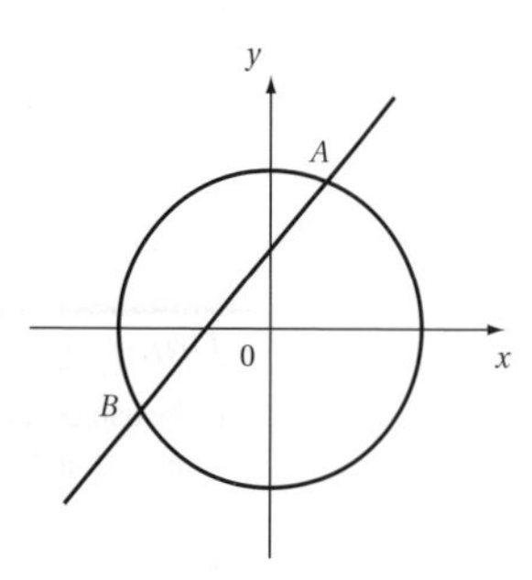

The line and the circle intersect at the points A and B.

a Show that the x-coordinates of A and B satisfy the equation $x^2 + 2x - 3 = 0$.

b Hence find the coordinates of A and B.

Working

a The equations are: $y = x + 2$ (1)

$x^2 + y^2 = 10$ (2)

You substitute the letter y in equation (2) with the expression $(x + 2)$ from equation (1).

So $x^2 + (x + 2)^2 = 10$

Expand the bracket: $x^2 + x^2 + 4x + 4 = 10$

Gather like terms: $2x^2 + 4x - 6 = 0$

Divide all terms by 2: $x^2 + 2x - 3 = 0$

This equation is satisfied by the x-coordinate of any intersection point of the line and circle. So the x-coordinates of A and B satisfy this equation, as required.

b Equation to solve: $x^2 + 2x - 3 = 0$

Factorise: $(x - 1)(x + 3) = 0$

So either $x - 1 = 0$ or $x + 3 = 0$.

The solution is $x = 1$ or $x = -3$.

Now substitute each of these values of x into the line equation $y = x + 2$ to find the y-coordinates of A and B.

When $x = 1$, $y = x + 2$
$= 1 + 2$
$= 3$

When $x = -3$, $y = x + 2$
$= -3 + 2$
$= -1$

So the points of intersection have coordinates $A(1,3)$ and $B(-3,-1)$.

> *Handy hint*
> Solving the equations $x^2 + y^2 = 10$ and $y = x + 2$ simultaneously gives the coordinates of where the line $y = x + 2$ intersects the circle $x^2 + y^2 = 10$.

> *Handy hint*
> Dividing all terms by 2 simplifies the quadratic without changing its solutions.

> *Handy hint*
> You could solve this equation by completing the square, but factorisation is simpler.

> *Handy hint*
> Do not use the non-linear equation to find y. This could produce invalid solutions.

Moving on to AS Level

At AS Maths you may be given a pair of simultaneous equations to solve where one equation is a mixture of squared and linear terms.

AS Level Example 8

Solve the simultaneous equations $x + 2y = 1$, $y^2 + 2x^2 - 3x + 1 = 0$.

Working

The equations are $x + 2y = 1$ (1)

$y^2 + 2x^2 - 3x + 1 = 0$ (2)

You need to make one of the letters in equation (1) the subject.

It is easier to make x the subject of equation (1) than y.

Equation (1): $x + 2y = 1$.

Subtract $2y$ from both sides: $x = 1 - 2y$

Now substitute x in equation (2) with the expression $(1 - 2y)$

$$y^2 + 2(1 - 2y)^2 - 3(1 - 2y) + 1 = 0$$

Expand $(1 - 2y)^2$: $y^2 + 2(1 - 4y + 4y^2) - 3(1 - 2y) + 1 = 0$

Remove the brackets: $y^2 + 2 - 8y + 8y^2 - 3 + 6y + 1 = 0$

Simplify: $9y^2 - 2y = 0$

Factorise: $y(9y - 2) = 0$

Solve: $y = 0$ or $9y - 2 = 0$

So $y = 0$ or $y = \frac{2}{9}$.

Now substitute each of these values of y into the *rearrangement* of equation (1) in turn to find the corresponding x-values.

Rearrangement of (1): $x = 1 - 2y$

When $y = 0$, $x = 1 - 2y$
$= 1 - 2(0)$
$= 1$

When $y = \frac{2}{9}$, $x = 1 - 2y$
$= 1 - 2\left(\frac{2}{9}\right)$
$= 1 - \frac{4}{9}$
$= \frac{5}{9}$

So the solutions are: $(x = 1, y = 0)$, $\left(x = \frac{5}{9}, y = \frac{2}{9}\right)$

Handy hint

Do **not** try to combine (1) and (2) in the same way you would for two linear equations.

Handy hint

Try to think ahead here – if you were to rearrange $x + 2y = 1$ to make y the subject, the answer would involve fractions.

Common error

$(1 - 2y)^2$ is **not** equal to $1 - 4y^2$ or even $1 + 4y^2$! Also, $-3(1 - 2y)$ does **not** simplify to $-3 - 6y$.

Handy hint

Do **not** use the non-linear equation to find the x values, as this may produce answers which are not solutions to both equations.

Common error

The solutions are **not** $(x = 0, y = 1)$, $\left(x = \frac{2}{9}, y = \frac{5}{9}\right)$

Take care to assign x and y their correct values and also to pair off the answers correctly.

Taking it further

C1 questions often require you to solve the type of equations shown in Example 8. The method of substitution is also used to find where a line intersects a circle when the circle is not centred at the origin.

5 Coordinate geometry 2

5.1 Transformations of graphs

What you should already know:

- how the graphs of $y = f(x) + a$, $y = f(x + a)$, $y = f(ax)$ and $y = af(x)$, where a is a constant, are related to the graph of $y = f(x)$.

In this section you will learn:

- how to keep track of the important points on a curve as it undergoes one or more transformations.

Revisiting GCSE

In GCSE Maths, you met the terms translation, stretch, reflection and rotation to describe transformations of shapes and graphs. For example, the translation vector $\begin{pmatrix} 2 \\ -3 \end{pmatrix}$ means that any point is shifted 2 units to the right along the x-axis and 3 units down the y-axis. So, using this vector, the point $P(1,4)$ has *image* $P'(3,1)$.

Handy hint

You add the vectors:
$\begin{pmatrix} 1 \\ 4 \end{pmatrix} + \begin{pmatrix} 2 \\ -3 \end{pmatrix} = \begin{pmatrix} 1+2 \\ 4-3 \end{pmatrix} = \begin{pmatrix} 3 \\ 1 \end{pmatrix}$

Key point

You can apply a transformation to a general graph with equation $y = f(x)$.
Here are the rules for transformations, where O is the origin and a is a constant.
Given the graph of $y = f(x)$, the graph of

i $y = f(x) + a$ is a translation of $y = f(x)$ by the vector $\begin{pmatrix} 0 \\ a \end{pmatrix}$

ii $y = f(x + a)$ is a translation of $y = f(x)$ by the vector $\begin{pmatrix} -a \\ 0 \end{pmatrix}$

iii $y = af(x)$, where $a > 0$, is a stretch along the y-axis, from O, scale factor a

iv $y = f(ax)$, where $a > 0$, is a stretch along the x-axis, from O, scale factor $\frac{1}{a}$

v $y = -f(x)$ is a reflection of $y = f(x)$ in the x-axis

vi $y = f(-x)$ is a reflection of $y = f(x)$ in the y-axis.

GCSE A* Example 1

Diagram 1 shows the graph of $y = f(x)$. On one copy of the diagram, sketch the graph of

a $y = f(x + 3)$

b $y = -f(x)$.

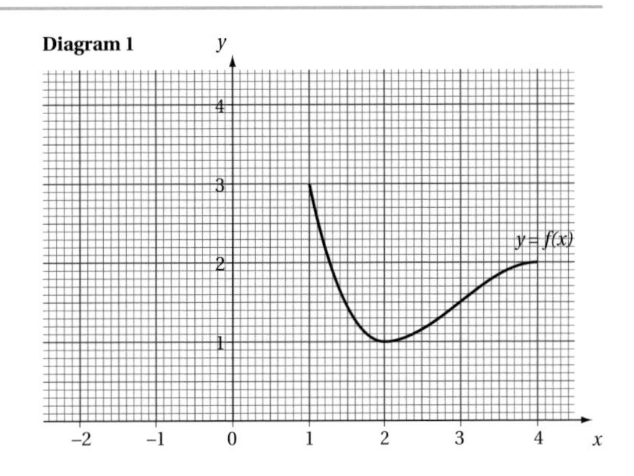

Working

a The graph of $y = f(x + 3)$ is a translation of $y = f(x)$ by the vector $\begin{pmatrix} -3 \\ 0 \end{pmatrix}$. See Graph A on Diagram 2.

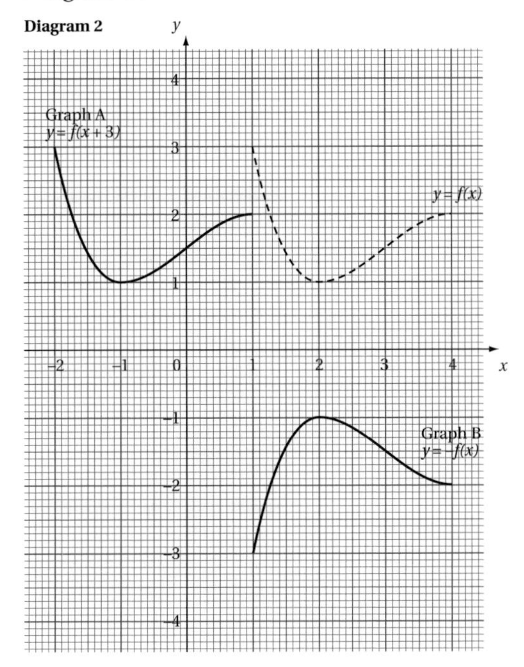

b The graph of $y = -f(x)$ is a reflection of $y = f(x)$ in the x-axis. See Graph B on Diagram 2.

Common error

$y = f(x + 3)$ is **not** a translation of $y = f(x)$ along the y-axis.

Moving on to AS Level

In AS Maths, you may need to keep track of several points on a graph as it undergoes a transformation.

AS Level Example 2

Figure 1 shows the graph of $y = f(x)$. The graph has a maximum point at $A(3,2)$ and a minimum point at $C(0,1)$. The graph crosses the x-axis at the point $B(6,0)$.

Sketch, on separate diagrams, the curve with equation

a $y = 2f(x)$

b $y = f(3x)$.

On each diagram show clearly the coordinates of the maximum and minimum points and where the curve crosses the x-axis.

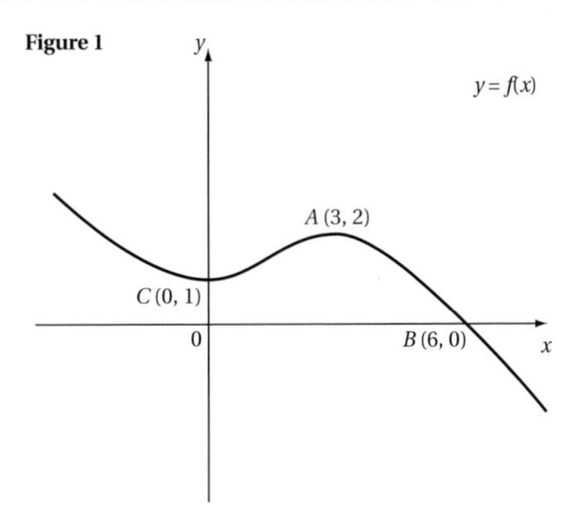

AS Alert!

You need to keep track of three points under each of these transformations. GCSE questions do not normally ask for this level of tracking.

Working

a You can sketch the graph of $y = 2f(x)$ by stretching the graph of $y = f(x)$ scale factor 2 along the y-axis from the origin O.

This means the y-coordinate of any point on $y = f(x)$ must be doubled.

Applying this transformation to the coordinates of the points A, B and C gives:

$$A(3,2) \xrightarrow{2 \times (y\text{-coordinate})} A'(3,4)$$

$$B(6,0) \xrightarrow{2 \times (y\text{-coordinate})} B'(6,0)$$

$$C(0,1) \xrightarrow{2 \times (y\text{-coordinate})} C'(0,2).$$

The transformed graph is shown on Figure 2.

Figure 2

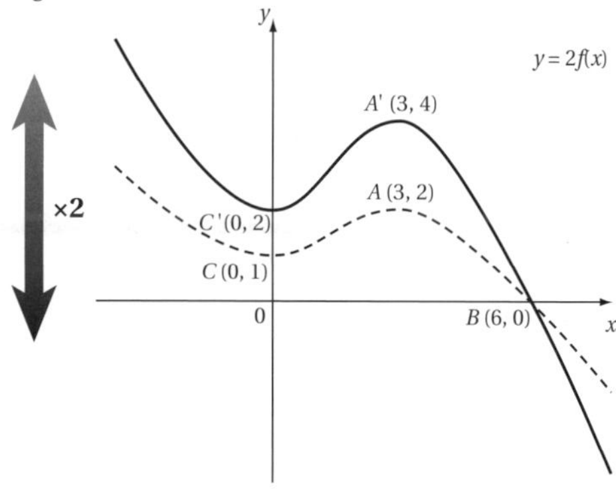

Handy hint

Under this transformation, points on the x-axis do not move. So $B' = B$.

Handy hint

You can sketch the dotted graph $y = f(x)$ to help visualise the stretch, but your final answer should just show the graph $y = 2f(x)$.

Common error

The transformed graph is **not** a translation of $y = f(x)$ along the y-axis.

Although there is some movement along the y-axis (e.g. $C \rightarrow C'$), the transformed graph has a different shape to that of $y = f(x)$.

A translation does not alter the shape of a graph.

b You must start with the graph $y = f(x)$

The graph of $y = f(3x)$ is a stretch along the x-axis of the graph $y = f(x)$, scale factor $\frac{1}{3}$ from the origin O.

This means to sketch the required graph, the x-coordinate of any point on $y = f(x)$ must be *divided* by 3.

Applying this transformation to the coordinates of the points A, B and C gives:

$$A(3,2) \xrightarrow{(x\text{-coordinate}) \div 3} A'(1,2)$$

$$B(6,0) \xrightarrow{(x\text{-coordinate}) \div 3} B'(2,0)$$

$$C(0,1) \xrightarrow{(x\text{-coordinate}) \div 3} C'(0,1).$$

The transformed graph is shown in Figure 3.

Common error

$y = f(3x)$ is **not** a stretch scale factor 3 along the x-axis.

Handy hint

Under this transformation, points on the y-axis do not move. So $C' = C$.

Figure 3

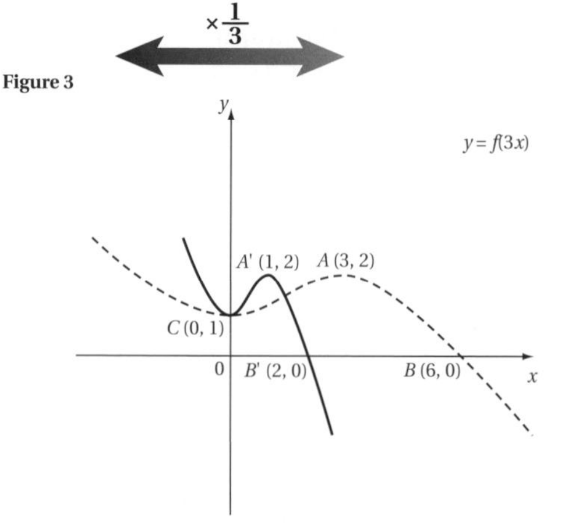

Taking it further

Transformations are used throughout AS Maths. For example, it is possible to describe the equation of a circle with centre at the point (3,4) as a transformation of a circle centred at the origin. You will also meet transformations applied to trigonometric graphs to help describe the behaviour of oscillating systems.

5.2 Sketching curves

What you should already know:

- how to sketch the graph of a simple quadratic or trigonometric curve such as $y = x^2 + 3$ or $y = \sin 2x$.

In this section you will learn:

- how to sketch more general quadratic functions such as $y = x^2 + 6x - 2$.
- how to apply transformations to sketch a reciprocal curve, such as $y = \frac{1}{x+4}$.

Revisiting GCSE

In GCSE Maths, you learnt how to sketch the graph of a quadratic which had undergone a simple transformation such as a translation or a stretch.

GCSE A Example 3

The dotted graph in the diagram shows the curve $y = x^2$.

On a copy of the diagram, sketch the graph $y = (x - 3)^2$.

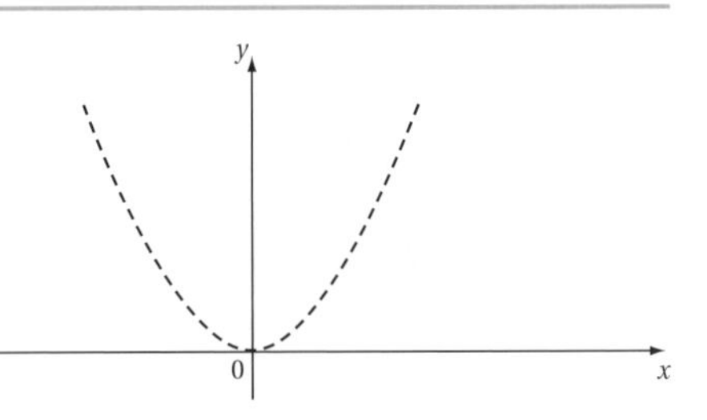

Working

The graph $y = (x - 3)^2$ is a translation of $y = x^2$ by the vector $\begin{pmatrix}3\\0\end{pmatrix}$.

So you shift the dotted graph 3 units to the right.

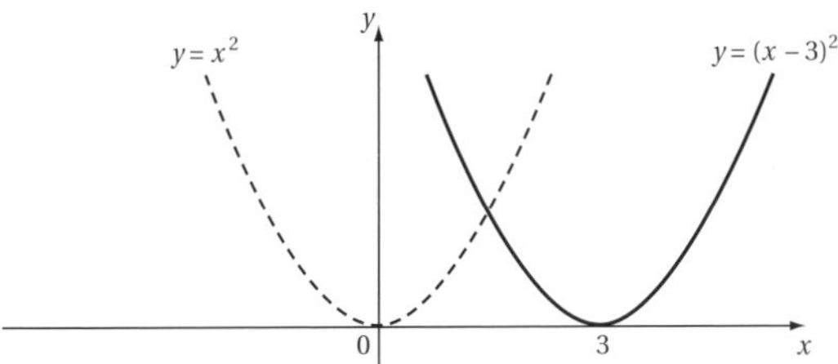

Handy hint

Refer to Section 5.1 for the rules of general transformations:
$(x - 3)^2 = (x + a)^2$ where $a = -3$, so the translation vector $\begin{pmatrix}-a\\0\end{pmatrix}$ is $\begin{pmatrix}3\\0\end{pmatrix}$.

Checkpoint

The sketch shows the curve $y = (x - 3)^2$ passing through the point (3,0).

This makes sense, because when $x = 3$, $y = (3 - 3)^2 = 0$.

Moving on to AS Level

In AS Maths you will be asked to sketch the graph of a general quadratic. One way of doing this is to express the quadratic in completed square form.

AS Level Example 4

a Express $x^2 + 4x + 9$ in the form $(x + p)^2 + q$, where p and q are constants.

b Hence, or otherwise, sketch the graph of $y = x^2 + 4x + 9$.

c Write down the equation of the line of symmetry of this curve.

AS Alert!

The phrase 'Hence, or otherwise' is used frequently; 'otherwise' means this curve can be sketched using another method, but 'hence' is suggesting you use the result of part **a**.

Working

a Express $x^2 + 4x + 9$ in completed square form:

$$x^2 + 4x + 9 = x^2 + 4x \qquad + 9$$
$$= (x + 2)^2 - 4 \quad + 9$$
$$= (x + 2)^2 + 5$$

So $x^2 + 4x + 9 = (x + 2)^2 + 5$.

b By part **a**, the graph you need to sketch is the same as the graph of $y = (x + 2)^2 + 5$.
To sketch this graph, you first translate the graph of $y = x^2$ by the vector $\begin{pmatrix} -2 \\ 0 \end{pmatrix}$.

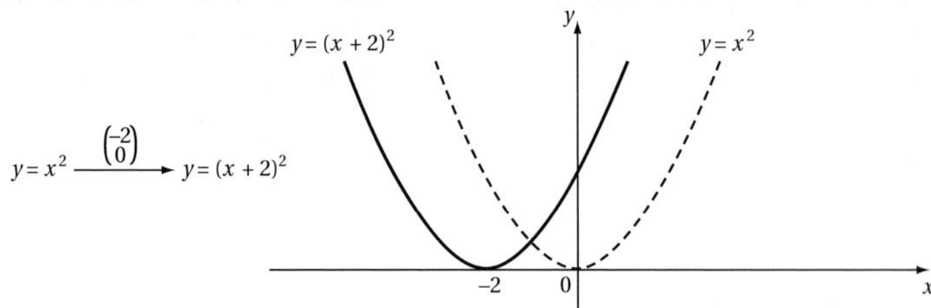

Next, you apply the translation vector $\begin{pmatrix} 0 \\ 5 \end{pmatrix}$ to the graph of $y = (x + 2)^2$.

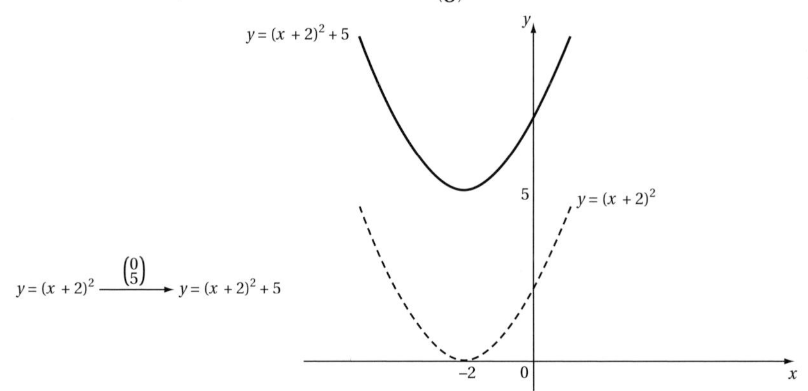

So the sketch of the graph $y = x^2 + 4x + 9$ is

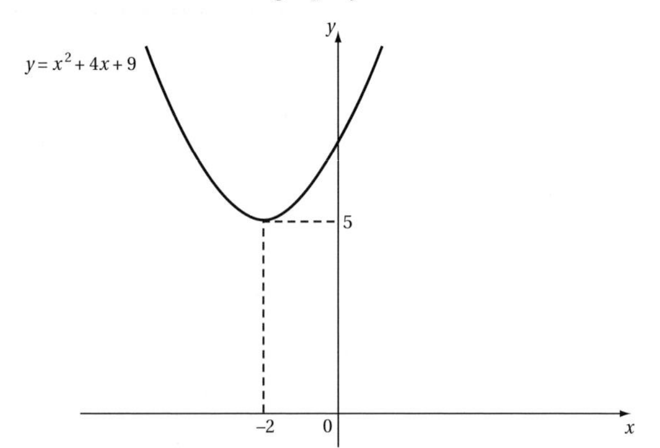

Handy hint

Show your final answer on a separate sketch, unless asked to do otherwise.

Handy hint

You can see that the final answer is simply the translation of $y = x^2$ by the vector $\begin{pmatrix} -2 \\ 5 \end{pmatrix}$. You can combine both movements into a single translation.

c The graph $y = x^2 + 4x + 9$ is symmetrical in the vertical line though the point $P(-2,5)$.

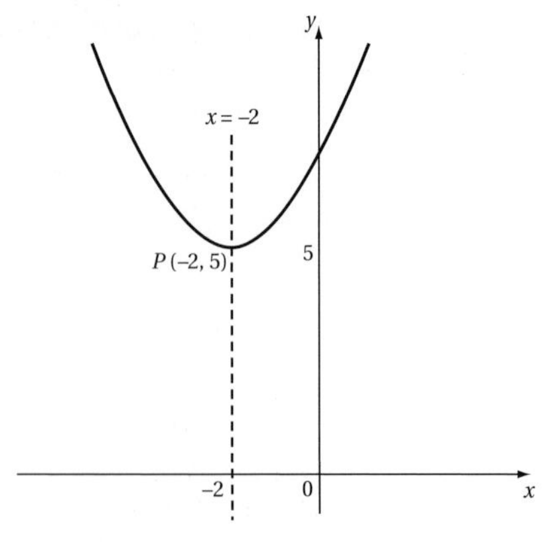

This line of symmetry has equation $x = -2$.

Handy hint

This means if you were to reflect this curve in the line $x = -2$, then the reflected graph would look identical to the original graph.

Point $P(-2,5)$ is the **minimum** point on this curve. This point is called the **vertex** of the curve.

Handy hint

As the question states, you must **write down** the equation of the line of symmetry.

Key point

The graph of $y = (x + p)^2 + q$, where p and q are constants, is a translation of $y = x^2$ by the vector $\begin{pmatrix} -p \\ q \end{pmatrix}$.
The coordinates of the minimum point (or *vertex*) of this curve are $(-p,q)$.
The line of symmetry of this curve has equation $x = -p$.

Checkpoint

Check this general result works when applied to the graph $y = (x - 3)^2$ in Example 3.

In AS Maths you may also have to deal with a graph that has an **asymptote**. An asymptote is a line which the graph approaches, but never touches, as x becomes very large and positive (or negative). For example, the asymptotes of the curve $y = \frac{1}{x}$ are the x-axis and the y-axis, as shown in Figure 1.

Figure 1

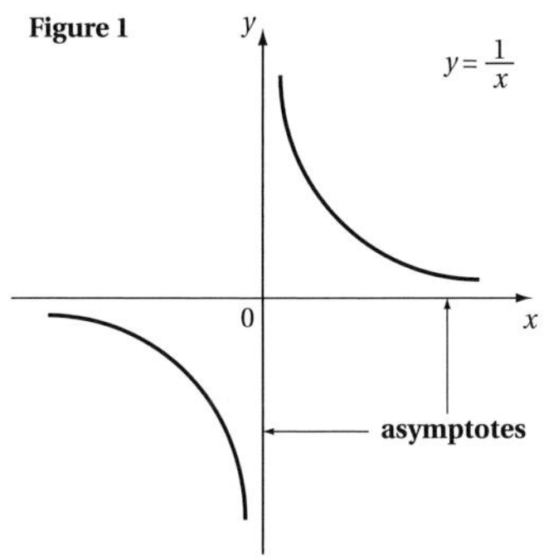

Handy hint

Remember that the graph of $y = \frac{1}{x}$ consists of two 'parts': one for $x > 0$ and one for $x < 0$.

The table of values explains why the graph of $y = \frac{1}{x}$ has this shape.

x	1	2	3	4
$y = \frac{1}{x}$	$\frac{1}{1} = 1$	$\frac{1}{2} = 0.5$	$\frac{1}{3} = 0.33\ldots$	$\frac{1}{4} = 0.25$

As x increases, the value of y decreases, and approaches zero. So the x-axis is an asymptote to this curve.

Handy hint

Similarly, as x approaches zero, $x > 0$, the value of y increases – check this for yourself.

AS Level Example 5

Sketch the graph of $y = \frac{2}{x} - 3$, stating the equations of the asymptotes of this curve.

Working

You start by stretching the graph of $y = \frac{1}{x}$ along the y-axis by scale factor 2 from the origin. You can visualise this transformation by applying the stretch to some points on the graph of $y = \frac{1}{x}$.

For example,

$$(1,1) \xrightarrow{2 \times (y\text{-coordinate})} (1,2)$$

$$(-1,-1) \xrightarrow{2 \times (y\text{-coordinate})} (-1,-2)$$

Figure 2 shows the dotted graph $y = \frac{1}{x}$ and the graph of $y = \frac{2}{x}$.

Figure 2

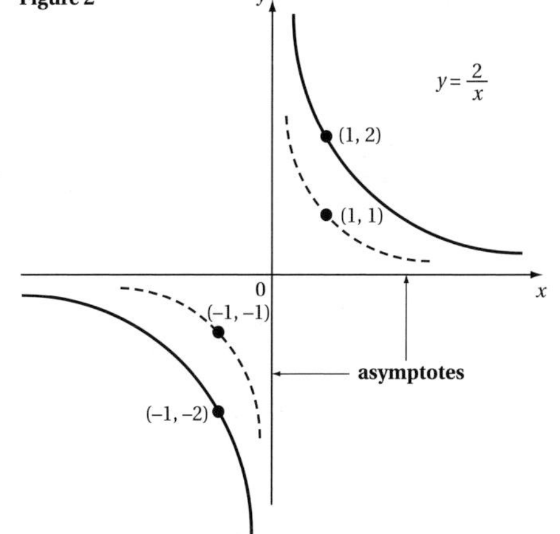

Handy hint

These asymptotes remain in their positions when stretching.

Next, you apply the translation vector $\begin{pmatrix} 0 \\ -3 \end{pmatrix}$ to the graph of $y = \frac{2}{x}$.

This means the graph of $y = \frac{2}{x}$ **and its asymptotes** must be shifted 3 units down the y-axis (see Figure 3).

Figure 3

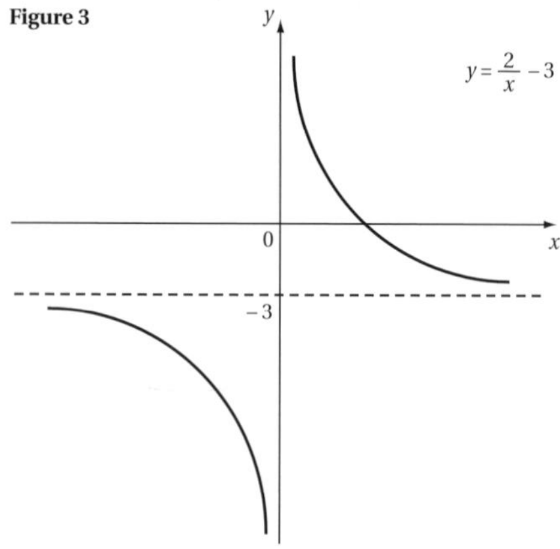

Handy hint

The asymptotes are part of the graph $y = \frac{2}{x}$ and so must shift 3 units down the y-axis.

The horizontal asymptote to the graph of $y = \frac{2}{x} - 3$ is the line $y = -3$.

The vertical asymptote to this graph is the line $x = 0$ (that is, the y-axis).

Key point

If a transformation involves a stretch and a translation then apply the stretch **first** and **then** the translation.

Handy hint

The order matters! In this example, if you apply the translation first and then the stretch, then you will have sketched the graph of $y = \frac{2}{x} - 6$. Check this for yourself.

Taking it further

You will be expected to sketch the graph of a general quadratic and identify its main features (e.g. the coordinates of its vertex). Transformations are one way of achieving this. You will also need to understand the significance of asymptotes when you study functions in more depth in unit C3.

5.3 Intersection points of graphs

What you should already know:

- how to find approximately where a line and a curve intersect by drawing their graphs.
- how to estimate the solution to an equation by drawing suitable graphs.

In this section you will learn:

- how to use algebra to find the exact coordinates where a line and curve intersect.
- how to use algebra to solve geometrical problems involving lines and curves.

Revisiting GCSE

In GCSE Maths, you may have found an approximate solution to an equation by using a graphical method.

GCSE A* Example 6

The diagram shows the curve $y = x^2 - 4x + 5$ and the line $y = 4 - x$. This curve and line intersect at the points P and Q.

An approximate solution of the equation $x^2 - 4x + 5 = 4 - x$ is $x = 0.4$.

a Explain how you can find this from the diagram.

b Use the diagram to find approximate solutions of the equation $x^2 - 3x + 1 = 0$.

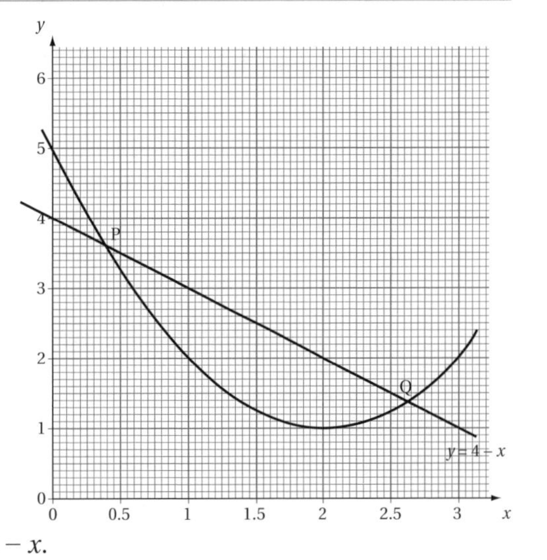

Working

a You can see that the solutions of the equation $x^2 - 4x + 5 = 4 - x$ are given by the x-coordinates of the points P and Q.

From the diagram, the x-coordinate of point P is approximately 0.4 and so this value must be an approximate solution of the equation $x^2 - 4x + 5 = 4 - x$.

b You need to rearrange the equation $x^2 - 4x + 5 = 4 - x$

Equation to rearrange: $x^2 - 4x + 5 = 4 - x$

Add x to both sides: $x^2 - 3x + 5 = 4$

Subtract 4 from both sides: $x^2 - 3x + 1 = 0$

Since $x^2 - 3x + 1 = 0$ is just a rearrangement of $x^2 - 4x + 5 = 4 - x$, the solutions of the equation $x^2 - 3x + 1 = 0$ are given by the x-coordinates of points P and Q.

You can read these solutions from the diagram. The equation $x^2 - 3x + 1 = 0$ has approximate solutions $x = 0.4$ and $x = 2.6$.

Key point

The x-coordinates of where a line and curve intersect correspond to the solutions to the equation found by putting the curve and line expressions equal to each other.

Moving on to AS Level

In AS Maths you are usually asked for exact values where two graphs intersect. This could involve answers in surd form. You may also be asked to use intersection points to solve a geometrical problem.

AS Level Example 7

In the diagram, the curve has equation $y = x^2 - 6x + 9$ and the line has equation $y = x + 3$. This line and curve intersect at points A and B, as shown. Points C and D on the x-axis are such that the lines AC and BD are vertical.

a Find the coordinates of the point where this curve intersects the x-axis.

b Find the coordinates of point A and point B.

c Show that the area of the trapezium $ABDC$ is 32.5 square units.

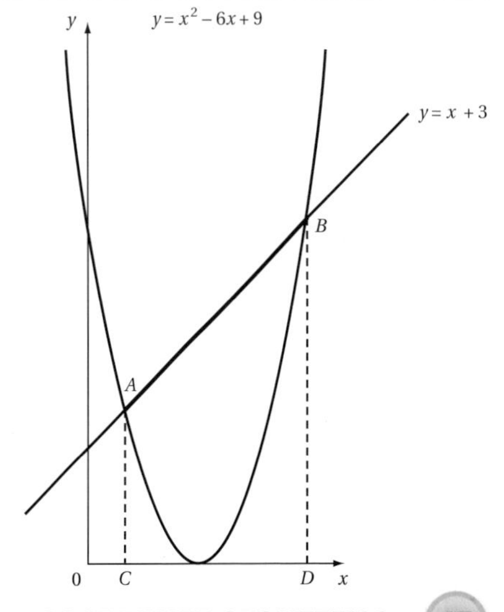

Working

a You can view the x-axis as the line with equation $y = 0$.

The equations are $y = 0$ (1)

$y = x^2 - 6x + 9$ (2)

You use the method of substitution, and replace the y term in equation (2) with the value 0 from equation (1).

Equation to solve: $x^2 - 6x + 9 = 0$

Factorise: $(x - 3)^2 = 0$

Solve: $x = 3$

So the curve intersects the x-axis at the single point (3,0).

> *Handy hint*
> See Section 3.5.

> *Checkpoint*
> In completed square form, the curve has equation $y = (x - 3)^2$ and so touches the x-axis at $x = 3$.

> *Checkpoint*
> The substitution method has produced a single point of intersection. This means the x-axis is a tangent to this curve.

b You apply the substitution method to the curve and line equations.

Equation to solve: $x^2 - 6x + 9 = x + 3$.

Gather terms to one side: $x^2 - 7x + 6 = 0$

Solve by factorising: $(x - 1)(x - 6) = 0$

So: $x - 1 = 0$ or $x - 6 = 0$

The solutions are $x = 1$ and $x = 6$

These solutions correspond to the x-coordinates of A and B.

So the x-coordinate of A is 1 and the x-coordinate of B is 6.

Substitute each x value into the line equation $y = x + 3$ to find the y-coordinates of A and B.

> *Handy hint*
> It's easier to use the line equation rather than the curve equation to find these y-coordinates (but the curve equation could be used as a check).

When $x = 1$, $y = x + 3$
$= 1 + 3$
$= 4$

When $x = 6$, $y = x + 3$
$= 6 + 3$
$= 9$

So A has coordinates (1,4) and B has coordinates (6,9).

c Use the coordinates of A and B to find the dimensions of the trapezium.

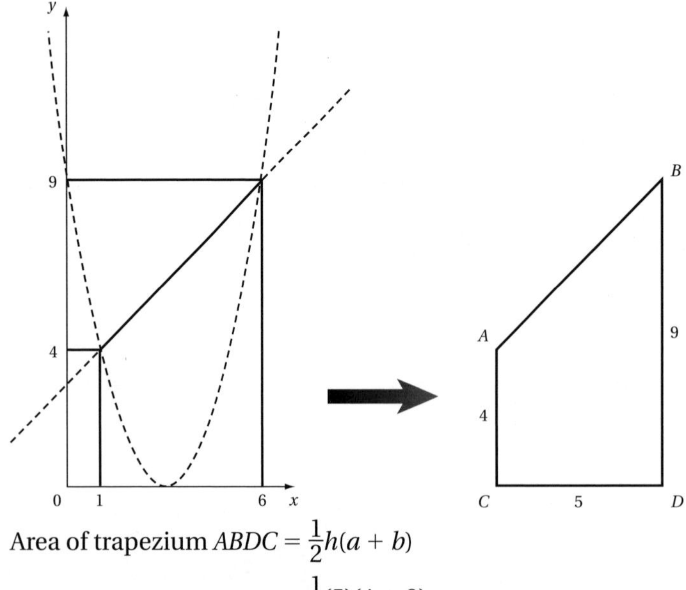

Area of trapezium $ABDC = \frac{1}{2}h(a + b)$

$= \frac{1}{2}(5)(4 + 9)$

$= \frac{1}{2} \times 65 = 32.5$ square units, as required

> *Handy hint*
> Use Area $= \frac{1}{2} \times h(a + b)$ where a and b are the lengths of the two parallel sides and h is the distance between them.

Key point

To find the x-coordinates of any points where a line and curve intersect, substitute for y and solve, where possible, the resulting equation.

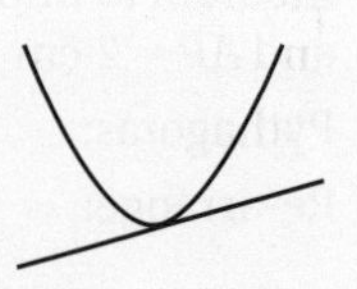

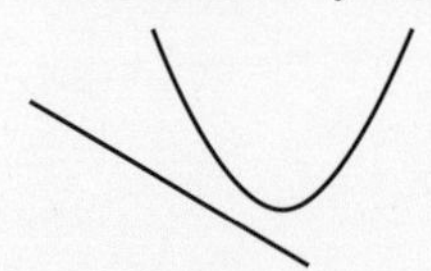

If there is only one solution to this equation, the line is a tangent to the curve

If there are no real solutions to this equation then the line does not intersect the curve.

Taking it further

You will need to understand the significance of intersection points of a line and curve when you study circles in unit C1 or unit C2. Also, in integration, you need to use the coordinates of intersection points to help calculate areas of regions.

5.4 Three circle theorems

What you should already know:

- how to apply various circle theorems to solve geometrical problems.

In this section you will learn:

- how to use the theorems to solve more complex problems.

Revisiting GCSE

In GCSE Maths, you would have met many different circle theorems. Most questions required you to find angles or lengths.

GCSE A* Example 8

The diagram shows two circles which touch each other externally.

The smaller circle has centre A and radius 2 cm.

The larger circle has centre B and radius 4 cm.

The line PQ is a tangent to both circles.

a Explain why AP and BQ are parallel.

b Given that $AQ = 6$ cm, find the area of triangle ABQ.
Give your answer to 1 decimal place.

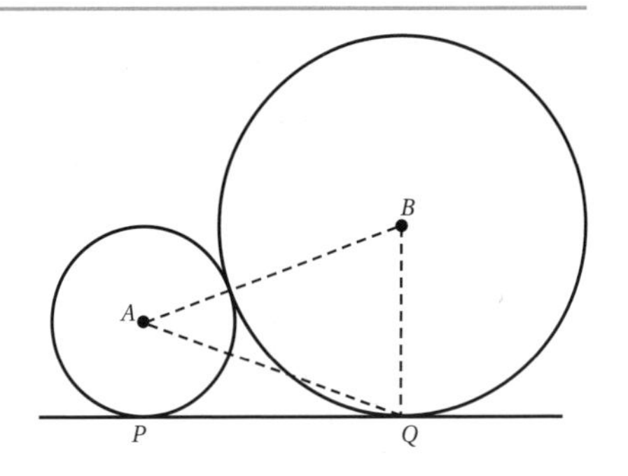

Working

a You need to use a circle theorem which involves tangents.

A radius makes a right-angle with a tangent.

So the line AP is perpendicular to the tangent PQ.

Similarly, the line BQ is perpendicular to the tangent PQ.

Hence AP and BQ are perpendicular to a common line and so are parallel to each other.

b You can use the formula Area $= \frac{1}{2}$ base $\times$ height,
where the base BQ of triangle ABQ has length 4 cm (the radius of the larger circle).

The height of triangle ABQ is equal to the distance PQ.

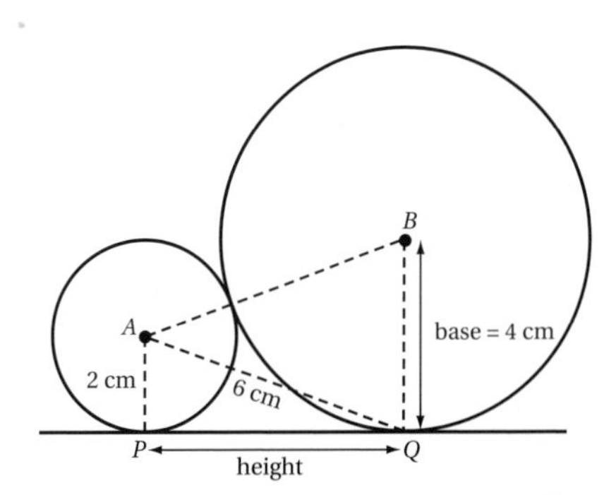

Triangle APQ is right-angled so you can use Pythagoras' theorem to find the length PQ, where $AQ = 6$ cm (as given) and $AP = 2$ cm (radius of smaller triangle).

Pythagoras: $AQ^2 = PQ^2 + AP^2$

Re-arrange: $PQ^2 = AQ^2 - AP^2$

$= 6^2 - 2^2$

$= 32$

Hence the height PQ = $\sqrt{32}$ cm.

So Area $= \frac{1}{2}$ base × height

$= \frac{1}{2} \times 4 \times \sqrt{32}$

$= 11.313...$ cm^2

Hence the area of triangle ABQ is 11.3 cm^2 to 1 decimal place.

Moving on to AS Level

In AS Maths, you only need to know **three** circle theorems. You may need to use them to find the centre or radius of a circle.

Key point

1 The angle in a semi-circle is a right angle.

2 The perpendicular from the centre to a chord bisects the chord.

3 The tangent to a circle is perpendicular to the radius at its point of contact.

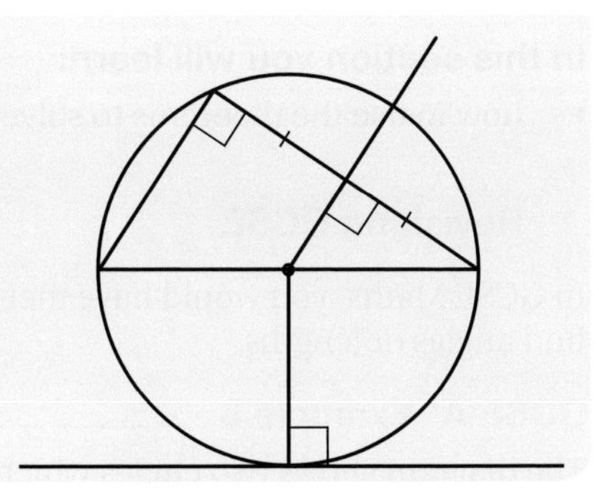

AS Level Example 9

The diagram shows a circle with centre $C(p,q)$, where $p > 0$.

The circle passes through the points $A(0,2)$ and $B(0,6)$.

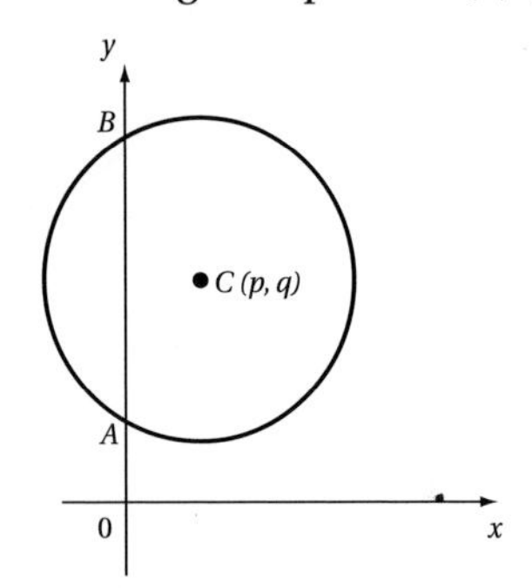

AS Alert!

Circles have centres given by coordinates and can be positioned anywhere, not just at the origin.

a Explain why $q = 4$.

b Given that angle $ACB = 90°$, find the value of p.

Working

a You need to recognise that AB is a chord of the circle and use circle theorem **2**.

The mid-point of $A(0,2)$ and $B(0,6)$ is the point $(0,4)$.

The horizontal line $y = q$, which passes through the centre C, bisects the vertical chord AB. Hence the line $y = q$ passes through the mid-point $(0,4)$ of AB.

So $q = 4$, as required.

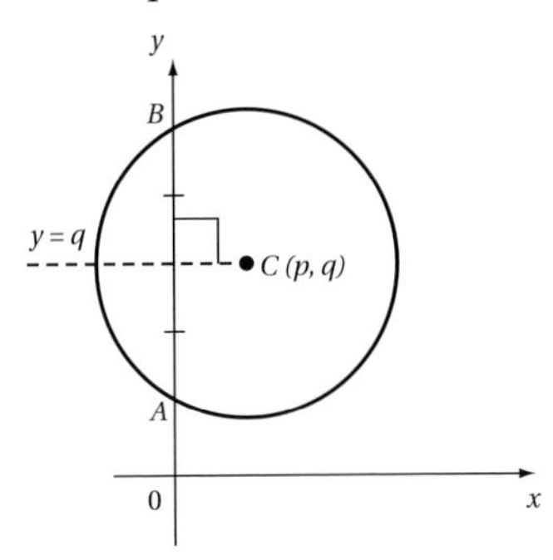

AS Alert!

The question did not use the word 'chord' – you have to see it for yourself!

b Since the lines AC and BC are perpendicular, you can use the result $Grad_{AC} \times Grad_{BC} = -1$.

$A(0,2)$, $C(p, 4)$ so $Grad_{AC} = \dfrac{4-2}{p-0} = \dfrac{2}{p}$

$B(0,6)$, $C(p, 4)$ so $Grad_{BC} = \dfrac{4-6}{p-0} = \dfrac{-2}{p}$

$Grad_{AC} \times Grad_{BC} = -1$ so $\dfrac{2}{p} \times \dfrac{-2}{p} = -1$

Multiply the fractions: $\dfrac{-4}{p^2} = -1$

Multiply both sides by p^2: $-4 = -p^2$

So: $p^2 = 4$

So $p = 2$ (as $p > 0$).

Handy hint

Check the details of the question – here $p > 0$ so you ignore the negative solution.

Taking it further

You will need to use these circle theorems in order to solve a variety of geometrical problems, for example, finding the equation of a tangent to a circle.

6 Trigonometry and triangles

6.1 Trigonometry and triangles

What you should already know:

- how to use the sine and cosine rules to find angles and lengths of sides of a triangle.

In this section you will learn:

- how to apply these rules to solve more complex problems.

Revisiting GCSE

In GCSE Maths, you learnt how to use trigonometric rules (usually called SOHCAHTOA) for right-angled triangles. You also met the sine and cosine rules, which can be applied to any triangle (including right-angled triangles!)

Key point

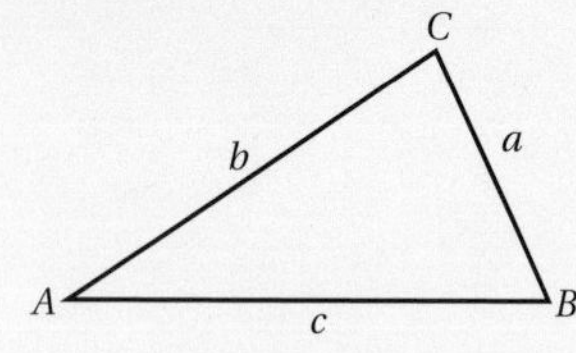

ABC is any triangle where $AB = c$, $AC = b$, $BC = a$ and $\hat{A}$ is the angle at vertex A etc.

sine rule: $\dfrac{a}{\sin \hat{A}} = \dfrac{b}{\sin \hat{B}} = \dfrac{c}{\sin \hat{C}}$.

By inverting each fraction, the sine rule can be written as

$$\frac{\sin \hat{A}}{a} = \frac{\sin \hat{B}}{b} = \frac{\sin \hat{C}}{c}.$$

cosine rule: $a^2 = b^2 + c^2 - 2bc \cos \hat{A}$.

which can be rearranged to give $\cos \hat{A} = \dfrac{b^2 + c^2 - a^2}{2bc}$.

Handy hint

The rule $\dfrac{\sin \hat{A}}{a} = \dfrac{\sin \hat{B}}{b} = \dfrac{\sin \hat{C}}{c}$ is useful for finding angles.

Handy hint

There are three cosine rules but they all have the same form. Write down the other two versions.

Handy hint

Make sure you can rearrange the cosine rule – the answer is not given in formula booklets.

Example 1 tells you which rule to apply.

GCSE A* Example 1

ABC is a triangle. Point D on BC is such that $BD = 11$ cm and $DC = 6$ cm. $\angle DAC = 25°$ and $\angle DCA = 35°$.

a Use the sine rule to find AD.

b Find angle ADB.

c Use the cosine rule to find AB.

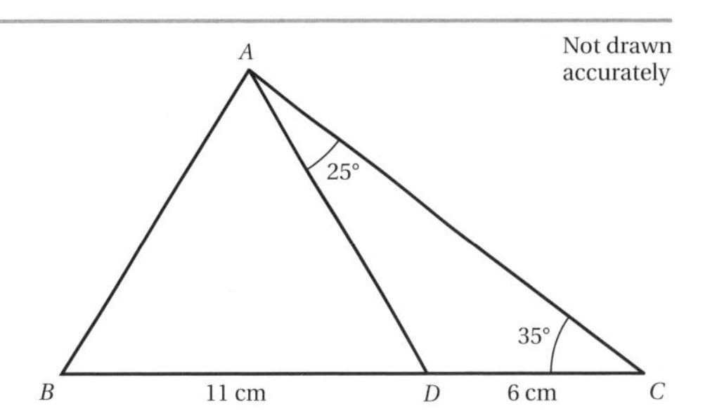

Working

a You need to apply the sine rule to triangle ADC.

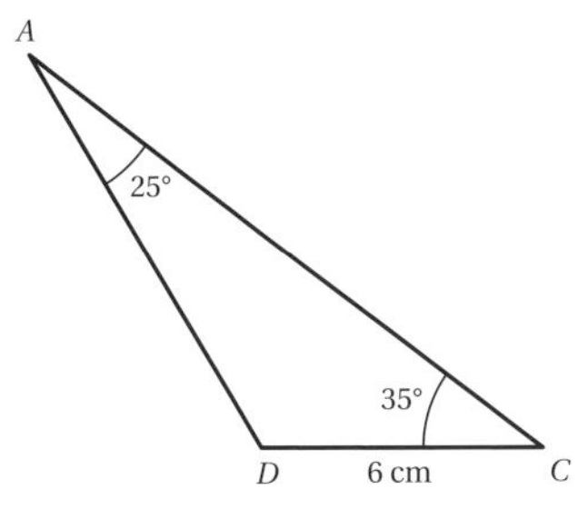

sine rule: $\frac{AD}{\sin 35°} = \frac{6}{\sin 25°}$

Multiply both sides by sin 35°: $AD = \left(\frac{6}{\sin 25°}\right) \times \sin 35°$

$= 8.14318\ldots$

So $AD = 8.14$ cm (2 decimal places).

b You can use the result that the sum of angles in a triangle equals 180°.

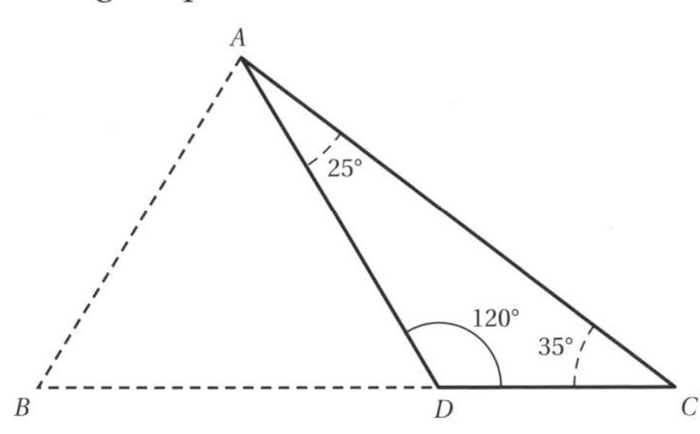

Angle $ADC = 180° - (25° + 35°)$

$= 180° - 60°$

$= 120°$

Hence angle $ADB = 180° - 120°$

$= 60°$

c You need to apply the cosine rule to triangle ABD.

cosine rule: $AB^2 = 11^2 + 8.14^2 - [2(11)(8.14)\cos 60°]$

$= 187.2596 - 89.54$

$= 97.7196$

So $AB = \sqrt{97.7196}$

$= 9.8853\ldots$

Hence $AB = 9.89$ cm (2 decimal places).

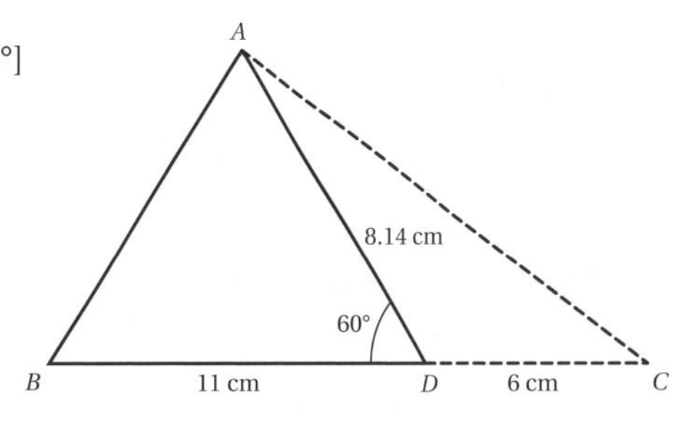

Handy hint

The vertices are not always A, B and C – you have to adapt the rule to fit the particular triangle.

Handy hint

Use the sine rule which has side lengths on the numerators.

Common error

AD is not $\left(\frac{\sin (35 \times 6)°}{\sin 25°}\right)$. Write sin 35° after the fraction to avoid this error.

Handy hint

In part **c**, use a square bracket to separate the two squared numbers from the rest of the calculation.

Moving on to AS Level

In AS Maths, you are rarely told which rule (sine or cosine) to use, and questions often involve problems set in a real-world situation.

AS Level Example 2

The diagram shows the plan of a large recreation park. The park ABC is in the shape of a sector of a circle with centre C and radius 1 km.

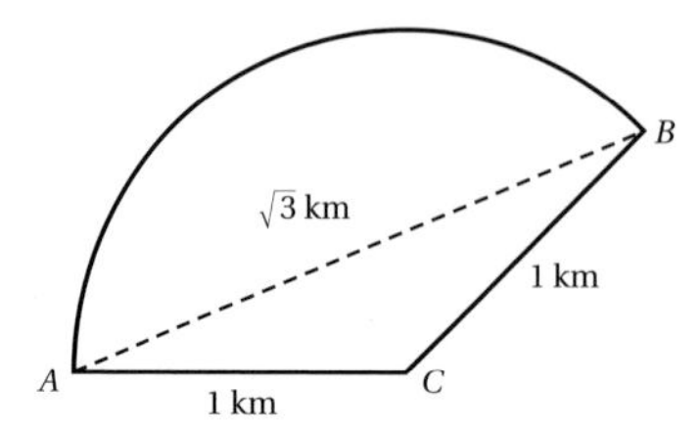

The length of the path AB is $\sqrt{3}$ km.

a Find the size of angle ACB.

An athlete runs along the perimeter of this park, starting and finishing at A.

b Find, in kilometres to 1 decimal place, the distance the athlete runs in one circuit.

> *AS Alert!*
> The question does not tell you which rule to use.

Working

a You need to use the cosine rule on triangle ABC.

cosine rule: $c^2 = a^2 + b^2 - 2ab\cos\hat{C}$.

Rearrange to make $\cos\hat{C}$ the subject.

Add $2ab\cos\hat{C}$ to both sides: $2ab\cos\hat{C} + c^2 = a^2 + b^2$

Subtract c^2 from both sides: $2ab\cos\hat{C} = a^2 + b^2 - c^2$

Divide both sides by $2ab$: $\cos\hat{C} = \dfrac{a^2 + b^2 - c^2}{2ab}$

Substitute in $a = 1, b = 1, c = \sqrt{3}$ $\cos\hat{C} = \dfrac{1^2 + 1^2 - (\sqrt{3})^2}{2(1)(1)}$

$= \dfrac{1 + 1 - 3}{2}$

$= -\dfrac{1}{2}$

Solve by using (cos⁻¹): $\hat{C} = \cos^{-1}\left(-\dfrac{1}{2}\right)$

$= 120°$

Hence angle $ACB = 120°$.

> *Handy hint*
> It's easy to make an error here – practise rearranging the cosine rule.

> *AS Alert!*
> The cosine of an angle can be negative. It can mean the angle is obtuse (i.e. between 90° and 180°).

b The circle has radius 1 km so the circumference of the circle is $2\pi \times 1 = 2\pi$ km.

So the length of the arc $AB = \left(\dfrac{120°}{360°}\right) \times 2\pi$

$= 2.0943\ldots$ km

(This answer uses the π button on a calculator.)

Hence the perimeter of the park $= 2.0943\ldots + 1 + 1$

$= 4.0943\ldots$

In one circuit the athlete runs 4.1 km (1 decimal place).

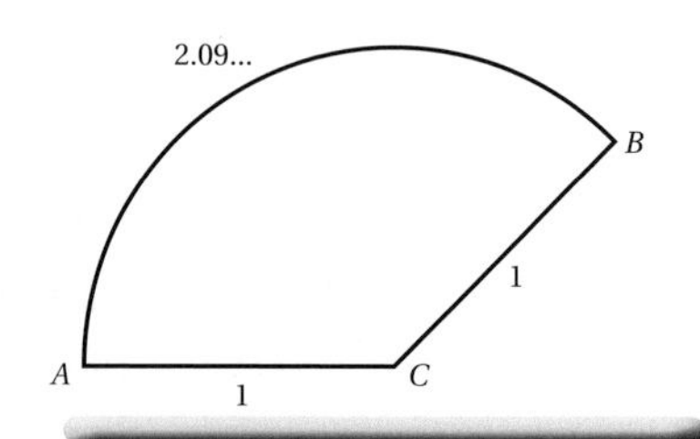

> *Handy hint*
> Remember to include the two straight edges AC and BC in your calculation.

Key point

In general, if you know three sides of a triangle, or two sides and the angle between them, use the cosine rule. Otherwise use the sine rule.

Taking it further

You will need to use the sine and cosine rules to answer different types of geometrical problems. You will be expected to know when to use each rule.

6.2 The area of any triangle

What you should already know:

- how to find the area of any triangle using two sides and the angle between them.

In this section you will learn:

- how to apply the area formula to a more complex problem.

Revisiting GCSE

In GCSE Maths, you used the formula $Area = \frac{1}{2}\ base \times height$ to find the area of a right-angled triangle. You may also have met the more general formula which finds the area of **any** triangle.

Key point

ABC is any triangle. $AB = c$, $AC = b$, $BC = a$.

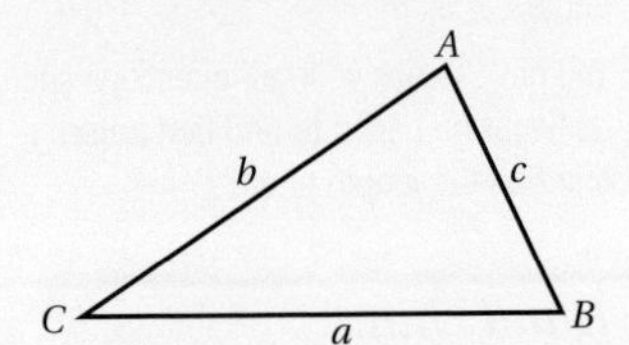

$Area = \frac{1}{2}ab \sin \hat{C}$, where $\hat{C}$ is the angle at vertex C.

The area of a triangle can be found if you know the angle between two known sides.

Handy hint

You can also write $Area = \frac{1}{2}ac \sin \hat{B}$ or $Area = \frac{1}{2}bc \sin \hat{A}$.

GCSE A Example 3

ABC is a triangle. $AB = 4$ cm, $AC = 7$ cm and angle $BAC = 78°$.

Calculate the area of the triangle. Give your answer to 3 significant figures.

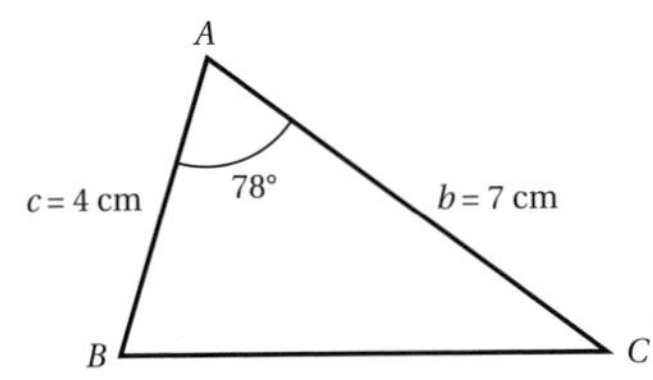

Working

You need to select the appropriate formula.

The given sides $c = 4$ cm and $b = 7$ cm enclose the given angle $\hat{A}$ so you use the rule $Area = \frac{1}{2}bc \sin \hat{A}$.

Substitute in the values.

$b = 7$, $c = 4$ and $\hat{A} = 78°$ so

$$Area = \frac{1}{2}bc \sin \hat{A}$$
$$= \frac{1}{2}(7)(4) \sin 78°$$
$$= 13.694\ldots$$

So the area of triangle ABC is 13.7 cm^2 (to 3 significant figures).

Moving on to AS Level

In AS Maths, questions involving areas of triangles are usually more involved and may require you to first find a length or an angle.

AS Level Example 4

The diagram shows a metal plate in which triangle ABC has been welded onto a semi-circular section. The semi-circle has diameter BC.

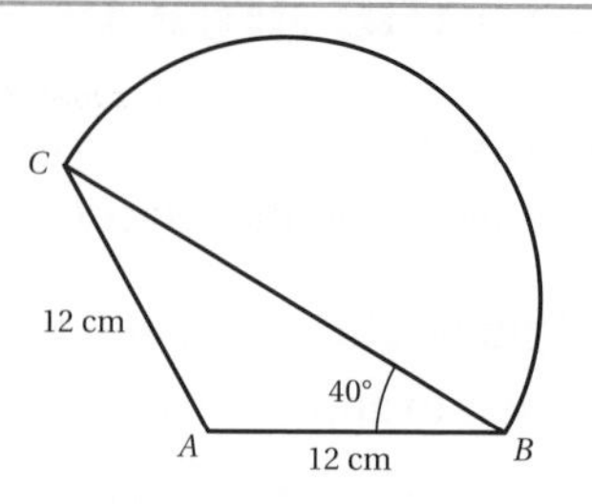

$AB = 12$ cm, $AC = 12$ cm and angle $CBA = 40°$.

a Show that the radius r of the semi-circle is 9.2 cm, to the nearest millimetre.

b Find, in cm^2 to 1 decimal place, the area of the plate.

Working

a You need to see that triangle ABC is isosceles.

$AC = AB$ so angle ACB = angle $CBA = 40°$

Hence angle $CAB = 180° - (40° + 40°)$

$= 100°$

Now use the sine rule to find side BC $(= a)$.

$$\frac{a}{\sin 100°} = \frac{12}{\sin 40°} \text{ so } a = \left(\frac{12}{\sin 40°}\right) \times \sin 100°$$

$$= 18.385\ldots$$

Hence $r = \frac{1}{2}BC = \frac{1}{2}(18.385\ldots)$

$= 9.192\ldots$

So $r = 9.2$ cm (1 decimal place), as required.

b The area of triangle ABC is $\frac{1}{2}bc \sin \hat{A} = \frac{1}{2}(12)(12) \sin 100°$

$= 70.906 \text{ cm}^2$ (3 d.p.)

The area of the semi-circle is $\frac{1}{2}\pi r^2 = \frac{1}{2}\pi(9.192)^2$

$= 132.721 \text{ cm}^2$ (3 d.p.)

Hence the total area $= 70.9061 + 132.721$

$= 203.627 \text{ cm}^2$

The area of the plate is 203.6 cm^2 (to 1 decimal place)

Handy hint

Look out for isosceles triangles in these types of questions.

AS Alert!

The question does not ask you to find angle CAB.

You have to ask yourself questions such as 'what do I need to find first before I can find the length of BC?'

Handy hint

You could also use the cosine rule to find BC.

Common error

The area of triangle ABC is not $\frac{1}{2} \times 12 \times 12$ because $\hat{A}$ is not a right-angle.

Handy hint

Avoid introducing rounding errors by using a sufficient number of decimal places in your working.

Taking it further

You will need to use the formula for the area of any triangle to solve various geometrical problems in unit C2. In particular, you will need it to find the areas of segments of circles.

6.3 Solving a trigonometric equation

What you should already know:

- the general shape and properties of the sine and cosine curves.
- how to find more solutions of a simple trigonometric equation, given one solution.

In this section you will learn:

- how to find all the solutions to more complex equations over a given range.

Revisiting GCSE

In GCSE Maths, you may have been given a solution to an equation involving sine or cosine and asked to find another solution to the same equation. One way to do this is to use the graphs of $y = \sin x$ or $y = \cos x$.

Key point

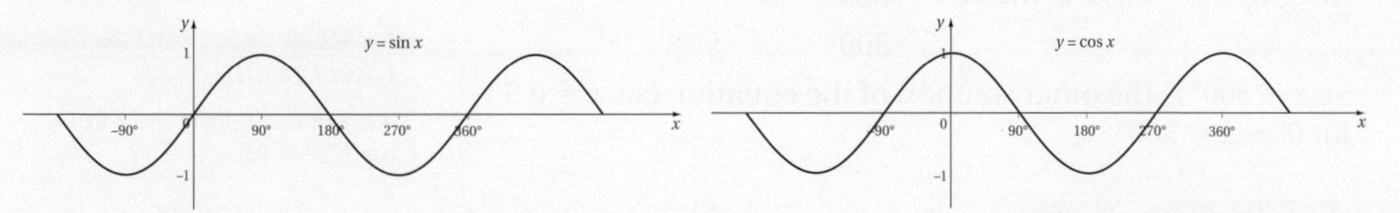

These graphs

1 are cyclic with period 360°
2 have a maximum value of 1 and a minimum value of -1
3 are defined for any positive or negative value of x.

Some important values on the sine graph are:
$\sin 0° = \sin 180° = \sin 360° = 0$, $\sin 90° = 1$, $\sin 270° = -1$.

Some important values on the cosine graph are:
$\cos 0° = \cos 360° = 1$, $\cos 90° = \cos 270° = 0$, $\cos 180° = -1$.

Handy hint

'period 360°' means a graph repeats every 360° – for example, $\sin 10° = \sin 370°$and, more generally, $\sin x° = \sin(360° + x°)$ for every angle x.

GCSE A Example 5

The diagram shows the graph of $y = \cos x$ for $0° \leqslant x \leqslant 360°$.

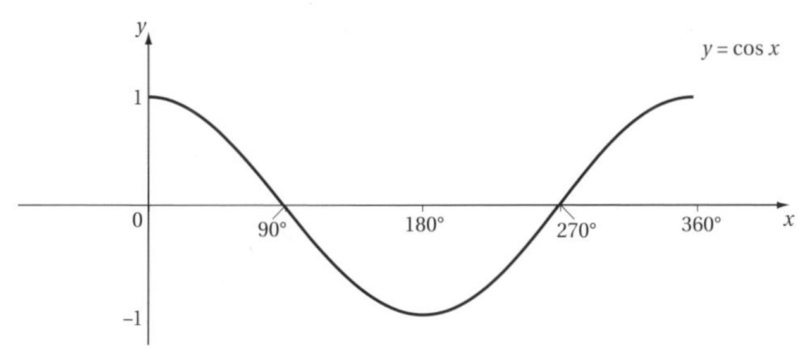

a Write down the solutions of the equation $\cos x = 0$ for $0° \leqslant x \leqslant 360°$.

One solution of the equation $\cos x = 0.5$ is $x = 60°$.

b Find another solution of the equation $\cos x = 0.5$ for $0° \leqslant x \leqslant 360°$.

Handy hint

You must find all the values of x between 0° and 360° for which $\cos x = 0$.

Working

a You can see from the graph that $\cos x = 0$ where the graph of $y = \cos x$ crosses the x-axis.

The solutions of $\cos x = 0$ for $0° \leqslant x \leqslant 360°$ are $x = 90°$ and $x = 270°$.

b One solution of the equation $\cos x = 0.5$ is $x = 60°$.

This means $\cos 60° = 0.5$.

So the line $y = 0.5$ intersects the graph of $y = \cos x$ at the point where $x = 60°$.

This intersection point is shown in Figure 1.

Handy hint

Intersection points correspond to solutions of equations – refer to Section 5.3.

Figure 1

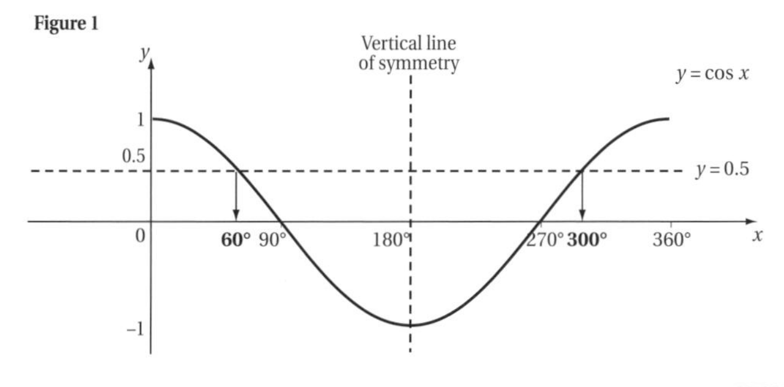

Checkpoint

This curve is symmetrical in the vertical line $x = 180°$.

So $x = (360° - 60°)$
$= 300°$

is another solution of this equation.

The next positive value of x where the line $y = 0.5$ intersects the graph $y = \cos x$ is where $x = 360° - 60°$

$$= 300°$$

So $x = 300°$ is the other solution of the equation $\cos x = 0.5$ for $0° \leqslant x \leqslant 360°$.

Checkpoint

Use your calculator to check that $\cos 300° = 0.5$.

Moving on to AS Level

In AS Maths, you will need to be able to solve more complicated equations without first being given a solution to work with.

AS Level Example 6

Solve the equation $3 \sin x + 2 = 3$ for $0° \leqslant x \leqslant 360°$. Give each answer to 1 decimal place.

Handy hint

You need to re-express the equation in the form $\sin x = k$ where k is a constant.

Working

You first need to find the value of $\sin x$.

Given equation: $3 \sin x + 2 = 3$

Subtract 2 from both sides: $3 \sin x = 1$

Divide both sides by 3: $\sin x = \frac{1}{3}$

Find one solution by using (sin⁻¹): $x = \sin^{-1}\left(\frac{1}{3}\right)$

$= 19.471\ldots$

$= 19.5°$ (1 decimal place).

AS Alert!

You need to find one solution of the given equation using your calculator.

Common error

The answer is **not** $\sin^{-1}(1) \div 3$.

Brackets are essential here!

Now sketch the graph $y = \sin x$ for $0° \leqslant x \leqslant 360°$ and the line $y = \frac{1}{3}$. Indicate the solution of this equation that you found on your calculator (see Figure 2).

Figure 2

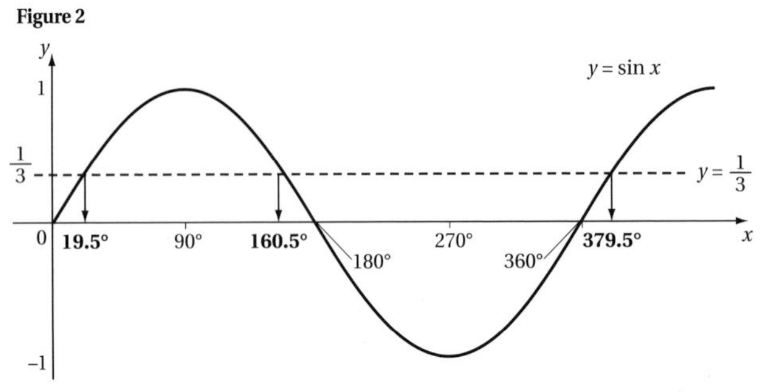

You can see that other values of x where this line and curve intersect are

$x = 180° - 19.5°$ $\quad$ $x = 360° + 19.5°$

$= 160.5°$ $\quad$ $= 379.5°$

The value $x = 379.5°$ does not lie in the range $0° \leqslant x \leqslant 360°$ so it must be ignored.

Hence, to 1 decimal place, $x = 19.5°$ and $x = 160.5°$ are the solutions of the equation $3 \sin x + 2 = 3$ for $0° \leqslant x \leqslant 360°$.

Handy hint

Take care to include only those values in the given range, in this case $0° \leqslant x \leqslant 360°$.

Taking it further

In unit C2, you will need to solve trigonometric equations such as $2 \sin (2x + 30°) = 1$ for $0° \leqslant x \leqslant 360°$. The techniques in this section can be used to solve such equations.

7 Sequences and summations

7.1 Arithmetic sequences

What you should already know:

- how to find the nth term of a simple sequence such as 3, 7, 11, ...

In this section you will learn:

- how to answer questions which involve arithmetic sequences.
- how to use arithmetic sequences to solve real-world problems.

Revisiting GCSE

In GCSE Maths, you met sequences which had a simple pattern. You also learnt how to find an expression for the nth term of the sequence.

GCSE C Example 1

A sequence of numbers is shown. 6, 10, 14 ...

a Find an expression for the nth term of this sequence.

b Explain why 1001 will not be a term in this sequence.

Working

a You need to find the increase from one term to the next.

The terms increase by 4. $6 \xrightarrow{+4} 10 \xrightarrow{+4} 14$

So, using the same pattern, the term before 6 would have to be $6 - 4 = 2$. $2 \xleftarrow{-4} 6$

So the nth term is: $4n + 2$

Increase from term to term (4) — The term before the sequence starts (2)

> *Handy hint*
> You may have been taught a different method for finding the nth term – any method that works is fine!

> *Handy hint*
> Subtracting 4 from 6 helps you find the sequence formula, but the answer 2 is not a term in this sequence.

> *Checkpoint*
> e.g. when $n = 3$
> $4n + 2 = 4(3) + 2 = 14$,
> which equals the 3rd term in the sequence 6, 10, **14**, ...

b Since 4 and 2 are even numbers, you can see that $4n + 2$ will also be an even number, for any value of n.

As 1001 is an odd number, it cannot be a term in this sequence.

The sequence in Example 1 is called **arithmetic**.

The difference between *any* pair of consecutive terms is a constant (in this case 4).

Handy hint

'Consecutive terms' are terms next to each other.

Key point

A sequence is arithmetic if the difference between **any** pair of consecutive terms is a constant. This is called the common difference, d.

Handy hint

$d =$ 'any term minus the *previous* term'.

So, in Example 1, $d = 10 - 6 = 4$.

Moving on to AS Level

In AS Maths you will need to use a standard formula for finding the nth term of an arithmetic sequence. This formula is useful for proving more general results.

Key point

If an arithmetic sequence has a common difference d and first term a then the nth term of the sequence is $a + (n - 1)d$.

You can check that this formula works for the sequence in part **a**, where $d = 4$ and $a = 6$.

Using the formula, the nth term is $a + (n - 1)d = 6 + (n - 1) \times 4$

Expand the bracket: $= 6 + 4n - 4$

Simplify: $= 4n + 2$

Checkpoint

This answer agrees with that found in part **a** of Example 1.

You can see why this formula works by considering the first few terms of an arithmetic sequence with common difference d and first term a. Each term is found by adding d to the previous term.

	1st term $= a$	which can be written as $a + \mathbf{0} \times d$
add d to give	**2nd** term $= a + d$	which can be written as $a + \mathbf{1} \times d$
add d to give	**3rd** term $= (a + d) + d$	which can be written as $a + \mathbf{2}d$
add d to give	**4th** term $= (a + 2d) + d$	which can be written as $a + \mathbf{3}d$.

By comparing the numbers in bold in each line you can see that, in general, the nth term can be written as $a + (n - 1)d$.

Sequence questions in AS Maths often use a symbol such as u_n.

This is simply short hand for 'the nth term of a sequence'.

So, for the sequence 6 , 10 , 14, … in part **a**, you can write $u_3 = 14$ and, in general, $u_n = 4n + 2$.

Handy hint

Think of u_n as a single expression. On its own, the letter u does not mean anything.

AS Level Example 2

The nth term of the sequence −2, 3, 8, 13, … is u_n.

a Find the 20th term.

b Which term in this sequence has value 193?

c Find the first term which is greater than 155.

Working

a You need to find u_{20}.

You should start by finding an expression for u_n.

The first term is $a = -2$.

The common difference $d = 3 - (-2)$
$= 5$

So: $u_n = a + (n - 1)d$
$= (-2) + (n - 1)(5)$

Expand the bracket: $= (-2) + 5n - 5$

Simplify: $= 5n - 7$

Now substitute $n = 20$ into this formula to find u_{20}.

$u_{20} = 5(20) - 7$
$= 100 - 7$
$= 93$

So the 20th term has value 93.

Handy hint

Do not attempt to count up to the 20th term! This would take too long and you might make a mistake.

AS Alert!

The question did not tell you to find u_n, but this will help you with all parts of this question.

Handy hint

You are allowed to use the formula $u_n = a + (n - 1)d$ – you do not have to explain why it works.

b You need to solve the equation $u_n = 193$.

Equation to solve: $5n - 7 = 193$

Add 7 to both sides: $5n = 200$

Divide both sides by 5: $n = 40$

So the 40th term has value 193.

Common error

The answer is **not** found by calculating u_{193}.

The question asks for the position of the term whose value 193, not for the 193rd term of the sequence.

c You need to solve the inequality $u_n > 155$.

Inequality to solve: $5n - 7 > 155$

Add 7 to both sides: $5n > 162$

Divide both sides by 5: $n = 32.4$

So the first term which is greater than 155 is the 33rd term.

Common error

The required value of n is **not** 32.

Although 32.4 is closer to 32, you must round n up to 33 because the terms are increasing. For $n = 32$, $u_n < 155$, for $n = 33$, $u_n > 155$.

In AS Maths you may need to describe a real-world situation using sequence notation.

AS Level Example 3

Crime figures in the city of Metropolis were recorded each month. The figures formed an arithmetic sequence with common difference d.

In January of a particular year, the crime figure was 1575. In March of the same year the figure was 1485.

a Find the value of d.

b Calculate the crime figure for November of the same year.

Working

a You should use sequence notation.

Let u_n = number of reported crimes in the nth month, where January means $n = 1$, February means $n = 2$, etc.

The figures form an arithmetic sequence with first term $a = 1575$ and common difference d.

So $u_n = a + (n - 1)d$
$= 1575 + (n - 1)d$

AS Alert!

You may need to define the sequence for yourself.

The figure for March (the 3rd month of the year) is 1485 so $u_3 = 1485$ where $u_3 = 1575 + (3 - 1)d$.

Equation to solve: $1575 + 2d = 1485$

Subtract 1575 from both sides: $2d = -90$

Divide both sides by 2: $d = -45$

So $d = -45$

b You can use the value of d to find an expression for u_n.

So: $u_n = a + (n - 1)d.$

$= 1575 + (n - 1)(-45)$

November is the 11th month of the year, so you must find the value of u_{11}.

Substitute $n = 11$ into the sequence formula.

$u_n = 1575 + (n - 1)(-45)$

So: $u_{11} = 1575 + (11 - 1)(-45)$

$= 1575 - 450$

$= 1125$

So the crime figure for November of that year is 1125.

Common error

d is **not** 45. It is tempting to use common sense to find d (e.g. the figures changed by 90 over a 2 month period, so $d = 45$).

The terms are decreasing so d must be negative! The safest way to find d is to use algebra.

Handy hint

You need not simplify this expression if you are simply calculating one of the terms (in this case u_{11}).

AS Alert!

You need to be able to translate words into an equation.

Taking it further

You will meet different types of sequences in AS Maths (e.g. geometric). You will need to be confident in using the various formulae for sequences and applying them to solve practical problems.

7.2 Recurrence relations and sigma notation

What you should already know:

- how to work with arithmetic sequences and the notation u_n.

In this section you will learn:

- how to work out one term of a sequence using previous terms.
- how to use sigma notation to simplify sums.

Handy hint

Refer to Example 1 in Section 7.1.

The arithmetic sequence 6, 10, 14, 18… has nth term formula $4n + 2$.

This sequence can also be described in words: 'starting with 6, you add 4 to one term to work out the next term'.

You can see this produces the same sequence 6, 10, 14, 18 ...

Start: **6**

$6 \xrightarrow{+4} \mathbf{10}$

$10 \xrightarrow{+4} \mathbf{14}$

$14 \xrightarrow{+4} \mathbf{18}$...

You can express these words using the notation u_n, where u_n means the nth term of the sequence.

Words	Sequence notation
'starting with 6'	$u_1 = 6$
'you add 4 to one term to work out the next term'	$u_n + 4 = u_{n+1}$

The formula $u_n + 4 = u_{n+1}$ is called a **recurrence relation** (or an iterative formula).

Handy hint

u_{n+1} is the $(n + 1)$th term. Think of u_n as the current term and u_{n+1} as the next term.

Key point

A **recurrence relation** is a formula for working out u_{n+1} given earlier terms in the sequence (such as u_n). A starting value is needed in order to work out the values of the sequence.

Recurrence relations enable you to describe complicated sequences in a simple way.

AS Level Example 4

A sequence is defined by: $u_1 = 3$

$u_{n+1} = 2u_n + 1.$

a Find the value of u_2

b Find the sum of the first four terms of this sequence

Handy hint

It is usual for the formula to have u_{n+1} on the left-hand side of the equation.

Working

a The starting value is $u_1 = 3$.

You need to use this value and the formula $u_{n+1} = 2u_n + 1$ to find u_2.

Apply the formula $u_{n+1} = 2u_n + 1$ when $n = 1$:

$u_{1+1} = 2u_1 + 1$

Now substitute u_1 with 3 so: $u_2 = 2(3) + 1$

$= 6 + 1$

So: $u_2 = 7$

Handy hint

In words, $u_{n+1} = 2u_n + 1$ means to find the next term u_{n+1}, double the current term u_n and then add 1.

b You need to first calculate the 3rd and 4th terms and then add together all four terms.

To find u_3, apply the formula $u_{n+1} = 2u_n + 1$ when $n = 2$

$u_{2+1} = 2u_2 + 1$

Now substitute u_2 with 7 so: $u_3 = 2(7) + 1$

$= 14 + 1$

So: $u_3 = 15$

Similarly, to find u_4, apply the formula $u_{n+1} = 2u_n + 1$ when $n = 3$: $u_{3+1} = 2u_3 + 1.$

Now substitute u_3 with 15 so: $u_4 = 2(15) + 1$

$= 30 + 1.$

So $u_4 = 31$

The first four terms of this sequence are $u_1 = 3$, $u_2 = 7$, $u_3 = 15$ and $u_4 = 31$.

So the sum of the first four terms is

$u_1 + u_2 + u_3 + u_4 = 3 + 7 + 15 + 31$

$= 56$

Note that it would not be possible to describe the sequence 3, 7, 15, 31, …using a simple expression such as $u_n = an + b$, or even using a quadratic expression $u_n = an^2 + bn + c$ for constants a, b and c.

In Example 4, you were asked to find the sum $u_1 + u_2 + u_3 + u_4$.

This sum can be written more compactly using the *summation* symbol $\sum$.

So, for example, $u_1 + u_2 + u_3 + u_4$ can be written as $\sum_{n=1}^{n=4} u_n$

You interpret $\sum_{n=1}^{n=4} u_n$ as meaning

- n takes each of the values 1, 2, 3, and 4 in turn
- for each value of n, you write down the term u_n
- add together the terms you have written down.

Handy hint

$\sum$ is the Greek uppercase letter **sigma**.

AS Level Example 5

The nth term of a sequence is given by $u_n = 2n + 1$.

Evaluate $\sum_{n=1}^{4} u_n$.

> *Handy hint*
>
> 'Evaluate' means 'find the numerical value of'. The 4 at the top of the sigma symbol means the final value of n is 4.

Working

You use the formula $u_n = 2n + 1$ and write out all the terms in the sum $\sum_{n=1}^{4} u_n$.

$$\sum_{n=1}^{4} u_n = \sum_{n=1}^{4} (2n + 1)$$
$$= (2 \times \mathbf{1} + 1) + (2 \times \mathbf{2} + 1) + (2 \times \mathbf{3} + 1) + (2 \times \mathbf{4} + 1)$$
$$= 3 + 5 + 7 + 9$$
$$= 24$$

So $\sum_{n=1}^{4} u_n = 24$

> *Handy hint*
>
> The numbers in bold are the values taken by n.

> *Handy hint*
>
> The answer can also be written as $\sum_{n=1}^{4} (2n + 1) = 24$, since $u_n = 2n + 1$.

Key point

If u_n is the nth term of any sequence then $\sum_{n=1}^{N} u_n$ is the sum of the first N terms of this sequence.

$$\sum_{n=1}^{N} u_n = u_1 + u_2 + u_3 + \ldots + u_N.$$

> *Handy hint*
>
> The starting value underneath the sigma notation need not be 1.
>
> e.g. $\sum_{n=3}^{5} u_n$ means $u_3 + u_4 + u_5$.
>
> Always look carefully at the starting value.

Taking it further

You will use recurrence relations as a way of describing sequences where it is difficult to find the nth-term formula. One of the most famous examples is the Fibonacci sequence 1, 1, 2, 3, 5, 8, …

8 Introducing differentiation

8.1 Estimating the gradient of a curve

What you should already know:

- how to find the gradient of a line passing through two points $P(x_1,y_1)$ and $Q(x_2,y_2)$.

In this section you will learn:

- how to define the gradient of a **curve** at any point.

Revisiting GCSE

You already know how to find the gradient of a line which passes through the points $P(x_1,y_1)$ and $Q(x_2,y_2)$.

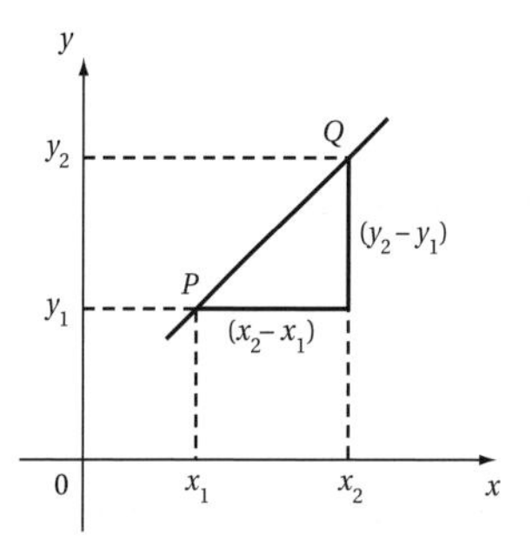

Key point

$$Grad_{PQ} = \frac{y_2 - y_1}{x_2 - x_1}$$

Clearly, the gradient of a line, once known, will never change.

Handy hint

$Grad_{PQ}$ means the gradient of the line PQ.

Think of $\frac{y_2 - y_1}{x_2 - x_1}$ as the 'change in y divided by the change in x'.

Moving on to AS Level

In AS Maths you will need to know how to find the gradient of a **curve**. Because it is not a straight line, the gradient of a curve will not be the same at all its points.

If you imagine walking up a hill, you will find some parts of your climb easier than others, because the hill is steeper in some places than others.

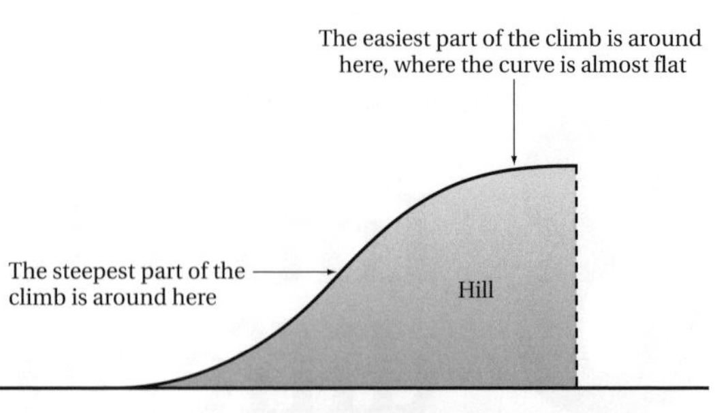

In order to understand precisely what is meant by the gradient of a curve at any point, consider 'zooming in' on the point $P(1,1)$ on the curve $y = x^2$.

> *Checkpoint*
>
> The y-coordinate of P is 1 because when $x = 1$, $y = 1^2 = 1$.

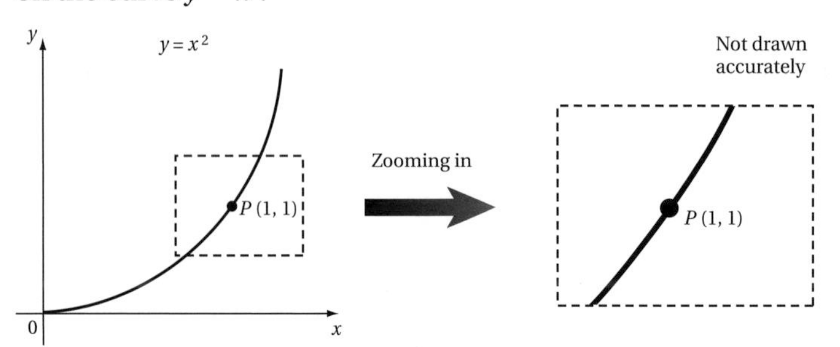

As you continuously zoom in on point P, the curve begins to look more and more like a straight line. This line is called the **tangent** to the curve at the point P.

Key point

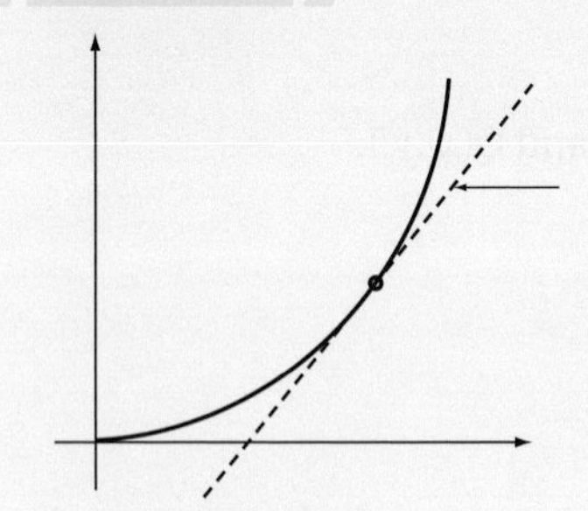

The tangent to a curve at point P is the straight line which the curve resembles as you zoom in on P.

The gradient of the curve at P is the gradient of this tangent.

> *Handy hint*
>
> A tangent to a curve at P just touches the curve at P. The tangent may intersect the curve at other points 'far away' from P.

AS Level Example 1

The diagram shows the curve $y = x^2$ which passes through the points $P(1,1)$, $A(2,4)$ and B, where the x-coordinate of B is 1.5.

a Find the gradient of the line

  i PA **ii** PB.

b State, with a reason, which of these two gradients gives the better estimate for the gradient of this curve at point P.

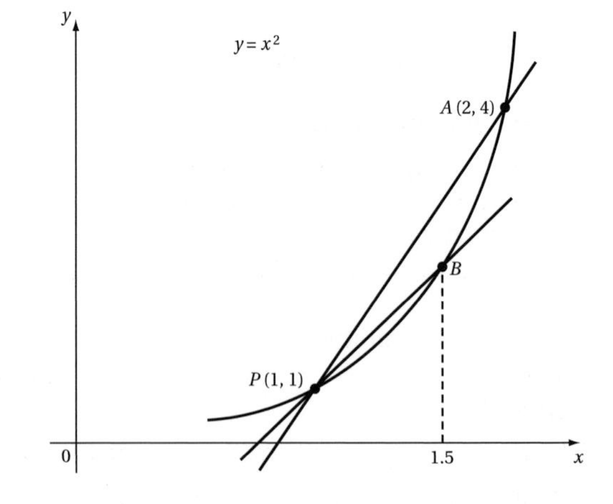

Working

a **i** You use the formula $Grad_{PA} = \frac{y_2 - y_1}{x_2 - x_1}$.

$P(1,1), A(2,4)$ so $Grad_{PA} = \frac{4 - 1}{2 - 1}$

$= \frac{3}{1}$

Hence $Grad_{PA} = 3$.

ii You first need to find the y-coordinate of B by substituting $x = 1.5$ into the equation $y = x^2$.

$x = 1.5$ so $y = x^2$

$= 1.5^2$

$= 2.25$

Hence the y-coordinate of B is 2.25.

So, using $P(1,1)$, $B(1.5,2.25)$, $Grad_{PB} = \frac{2.25 - 1}{1.5 - 1}$

$= \frac{1.25}{0.5}$

$= 2.5$

So $Grad_{PB} = 2.5$.

b The better estimate for the gradient of the curve at point P is $Grad_{PB} = 2.5$.
This is because point B is closer to P than point A.

Key point

The tangent to a curve at point P is approximated by the line PQ, where Q is a point on the curve which is close to P.
The closer Q is to P, the better the approximation.
The gradient of the curve at P is approximately $Grad_{PQ}$.

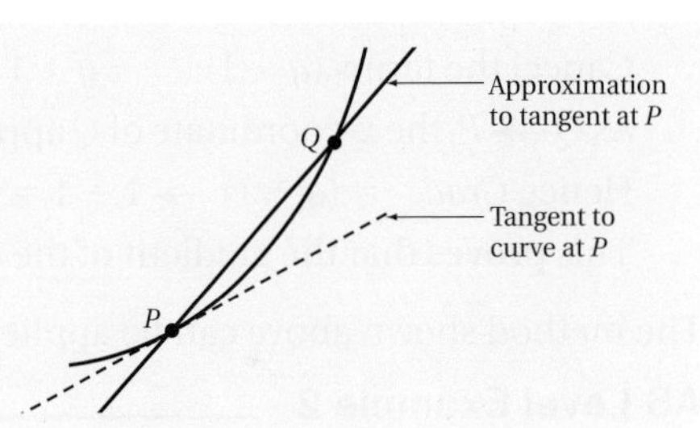

Taking it further

During your AS course you will learn how to find the **exact** gradient of a curve at any point. One way of doing this is shown in 8.2.

8.2 Differentiation

What you should already know:

- how to approximate the gradient of curve at a point P by calculating $Grad_{PQ}$, where Q is a point on the curve which is close to P.

In this section you will learn:

- how to find the exact gradient of a curve at any point.

You saw in Section 8.1 that the gradient of the curve at a point P is approximately equal to $Grad_{PQ}$, where Q is a point on the curve 'close' to P.
The closer Q is to P, the better the approximation.
The table on the next page shows the gradient of PQ where Q is a point on the curve $y = x^2$ which approaches the fixed point $P(1,1)$.
Fixed point: $P(1,1)$

x-coordinate of Q	y-coordinate of Q	$Grad_{PQ}$
1.1	$1.1^2 = 1.21$	$= \dfrac{1.21-1}{1.1-1} = 2.1$
1.01	$1.01^2 = 1.0201$	$= \dfrac{1.0201-1}{1.01-1} = 2.01$
1.001	$1.001^2 = 1.002001$	$= \dfrac{1.002001-1}{1.001-1} = 2.001$

You can see that as Q approaches P, $Grad_{PQ}$ approaches the value 2 or, written mathematically, as $Q \to P$, $Grad_{PQ} \to 2$.

This suggests that the **exact** gradient of the curve $y = x^2$ at the point $P(1,1)$ is 2.

Handy hint

The symbol $\to$ means 'approaches' or 'gets closer to'.

In order to be sure that this exact gradient is 2, you need to consider $Grad_{PQ}$ where Q is a **general** point on the curve, rather than a specific one such as $Q(1.1,1.21)$.

In the diagram, the point $Q(q,q^2)$ is any point on the curve $y = x^2$ which is close to the fixed point $P(1,1)$. The tangent to the curve at P is shown as a dotted line.

Using the coordinates of the points $P(1,1)$ and $Q(q,q^2)$.

$$Grad_{PQ} = \frac{q^2 - 1}{q - 1}$$

Factorise the numerator: $= \dfrac{(q-1)(q+1)}{q-1}$

Cancel the terms $(q - 1)$: $= q + 1$

As $Q \to P$, the x-coordinate of Q approaches the x-coordinate of P (i.e. $q \to 1$).

Hence $Grad_{PQ} = (q + 1) \to 1 + 1 = 2$.

This **proves** that the gradient of the curve $y = x^2$ at the point $P(1,1)$ is 2.

The method shown above can be applied to any point P on the curve $y = x^2$.

AS Level Example 2

Prove that the gradient of the curve $y = x^2$ at the point $P(x,x^2)$ is $2x$.

Working

You consider any point $Q(q,q^2)$ on the curve $y = x^2$ which is close to the general point $P(x,x^2)$. The tangent to the curve at P is shown as a dotted line.

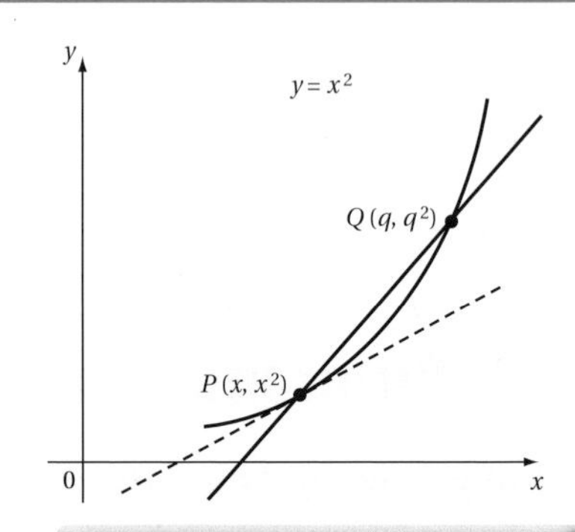

Using the coordinates of the points $P(x,x^2)$ and $Q(q,q^2)$,

$$Grad_{PQ} = \frac{q^2 - x^2}{q - x}$$

Factorise the numerator: $= \dfrac{(q-x)(q+x)}{q-x}$

Cancel the terms $(q - x)$: $= q + x$

As $Q \to P$, $q \to x$ and so $Grad_{PQ} = (q + x) \to x + x = 2x$.

Hence the gradient of the curve $y = x^2$ at the point $P(x,x^2)$ is $2x$.

Handy hint

Although P is a general point on the curve, you must think of it being fixed. It is Q that is moving towards P.

Key point

The process of finding an expression for the gradient of a curve at any point is called **differentiation from first principles**.

The notation $\frac{dy}{dx}$ is the gradient equation of a curve at any point.

Handy hint

Think of the letter d as meaning 'change' so that $\frac{dy}{dx}$ means 'change in y over change in x' i.e. a gradient.

So, Example 2 shows that for the curve $y = x^2$, $\frac{dy}{dx} = 2x$.

When you **differentiate** a curve, you are finding its gradient equation.

Handy hint

$\frac{dy}{dx}$ is pronounced 'dee y by dee x'.
However, $\frac{dy}{dx}$ is **not** a fraction, so you cannot cancel the 'd's!

AS Level Example 3

a Expand and then simplify $(q - x)(q^2 + qx + x^2)$.

b Hence show from first principles that for the curve $y = x^3$, $\frac{dy}{dx} = 3x^2$.

Working

a You expand the brackets:

$$(q - x)(q^2 + qx + x^2) = q(q^2 + qx + x^2) - x(q^2 + qx + x^2)$$
$$= q^3 + q^2x + qx^2 - xq^2 - x^2q - x^3$$

Simplify:
$$= q^3 + \cancel{q^2x} + \cancel{qx^2} - \cancel{xq^2} - \cancel{x^2q} - x^3$$
$$= q^3 - x^3$$

So $(q - x)(q^2 + qx + x^2) = q^3 - x^3$

b You consider any point $Q(q,q^3)$ on the curve $y = x^3$ which is close to the general point $P(x,x^3)$. The tangent to the curve at P is shown as a dotted line.

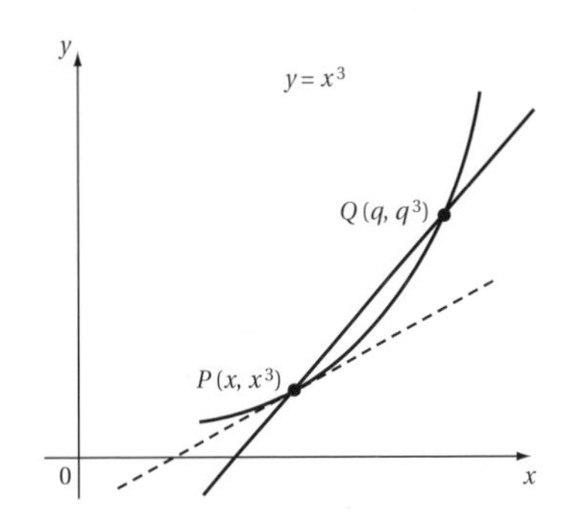

Using the coordinates of the points $P(x,x^3)$ and $Q(q,q^3)$,

$$Grad_{PQ} = \frac{q^3 - x^3}{q - x}$$

Factorise the numerator: $= \frac{(q - x)(q^2 + qx + x^2)}{q - x}$

Cancel the terms $(q - x)$: $= q^2 + qx + x^2$

As $Q \rightarrow P$, $q \rightarrow x$

and so $Grad_{PQ} = (q^2 + qx + x^2) \rightarrow (x^2 + xx + x^2)$
$$= x^2 + x^2 + x^2$$
$$= 3x^2$$

Hence the curve $y = x^3$ has gradient equation $\frac{dy}{dx} = 3x^2$, as required.

Handy hint

This is a 'Hence' question so make use of the result from part **a**.

You can use first principles to show similar results hold for other curves, such as $y = x^4$. Some curves and their gradient equations are shown in the table.

Curve equation	Gradient equation of curve
$y = x^2$	$\frac{dy}{dx} = 2x$
$y = x^3$	$\frac{dy}{dx} = 3x^2$
$y = x^4$	$\frac{dy}{dx} = 4x^3$

Handy hint

When you **differentiate** $y = x^4$ the answer is $\frac{dy}{dx} = 4x^3$.

You say the **derivative** of x^4 with respect to x is $4x^3$.

The results in the table are examples of the following rule.

Key point

If $y = x^n$, where n is any number, then $\frac{dy}{dx} = nx^{n-1}$

You can use this rule to differentiate more complicated equations **without** having to use first principles.

AS Level Example 4

Find $\frac{dy}{dx}$ for these equations.

a $y = 2x^3 + 4x^2$

b $y = \frac{1}{2}x^4 + 3x + 2$

Working

a You apply the $\frac{dy}{dx} = nx^{n-1}$ rule to each term in the equation.

Curve equation: $y = 2(x^3) + 4(x^2)$

Apply the rule to x^3 and x^2: $\frac{dy}{dx} = 2(3x^2) + 4(2x)$

Simplify: $= 6x^2 + 8x$

So: $\frac{dy}{dx} = 6x^2 + 8x$

Handy hint

You may find it helpful to put brackets around the x-terms.

Handy hint

Bring down the coefficients 2 and 4 in the curve equation.

b You can express $\frac{1}{2}x^4 + 3x + 2$ using powers of x.

The term $3x$ can be written as $3x^1$.

The constant 2 can be written as $2x^0$.

So, curve equation: $y = \frac{1}{2}(x^4) + 3(x^1) + 2(x^0)$

Apply the rule to each power of x: $\frac{dy}{dx} = \frac{1}{2}(4x^3) + 3(1x^0) + 2\,(0x^{-1})$

Gather the coefficients: $= \left(\frac{1}{2} \times 4\right)x^3 + (3 \times 1)x^0 + (2 \times 0)x^{-1}$

Simplify: $= 2x^3 + 3$

So: $\frac{dy}{dx} = 2x^3 + 3$

Handy hint

$a^0 = 1$ for any value of a.

Handy hint

Notice that the derivative of $3x + 2$ is 3, the coefficient of x.

You can use this shortcut when differentiating linear expressions.

Key point

If $y = mx$, where m is a constant, then $\frac{dy}{dx} = m$.

If $y = c$, where c is a constant, then $\frac{dy}{dx} = 0$.

So, if $y = mx + c$ then $\frac{dy}{dx} = m$.

Checkpoint

This result makes sense since the line $y = mx + c$ has gradient m.

You may need to simplify an equation before it can be differentiated.

AS Level Example 5

A curve has equation $y = (2x + 1)(x - 2)$.

a Show that $\frac{dy}{dx} = 4x - 3$.

b Find the gradient of this curve at the point $P(3,7)$.

Working

a You need to expand the brackets and simplify the equation *before* you can differentiate it.

Curve equation: $y = (2x + 1)(x - 2)$

Expand the brackets and simplify: $= 2x^2 - 3x - 2$

Apply the rule: $\frac{dy}{dx} = 2(2x) - 3$

So: $\frac{dy}{dx} = 4x - 3$, as required.

Handy hint

The derivative of $-3x - 2$ is -3 (see Key point).

b You substitute the x-coordinate of $P(3,7)$ into the equation $\frac{dy}{dx} = 4x - 3$.

When $x = 3$, $\frac{dy}{dx} = 4x - 3$

$= 4(3) - 3$

$= 9$

So the curve $y = (2x + 1)(x - 2)$ at the point $P(3,7)$ has gradient 9.

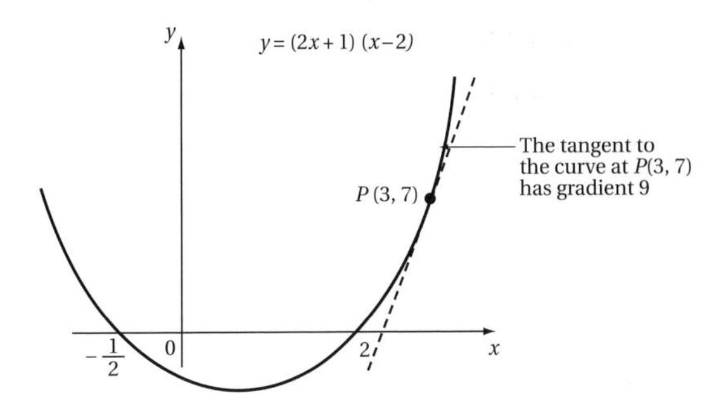

Key point

Working out the value of $\frac{dy}{dx}$ at a particular point P on the curve gives the gradient of the curve at point P. This is also equal to the gradient of the tangent at P.

Taking it further

You will learn how to use differentiation to find the equation of a tangent to a curve. Differentiation also has uses in practical problems such as maximising volumes of containers.

8.3 Applications of differentiation

What you should already know:

- how to differentiate a curve to find its gradient equation.
- how to calculate the gradient of a curve at any given point.

In this section you will learn:

- how to find the equation of the tangent and the normal to a curve at any point.
- how to locate stationary points on a curve.

In Section 8.2 you saw how to differentiate an equation and use the answer to find the gradient of the tangent to the curve at a specific point.

You can go one step further and find the equation of the tangent to a curve at a given point. The following result is helpful for this.

Key point

If a line L has gradient m and passes through the point $P(x_1,y_1)$ then an equation for L is $y - y_1 = m(x - x_1)$.

The diagram shows why this result is true.

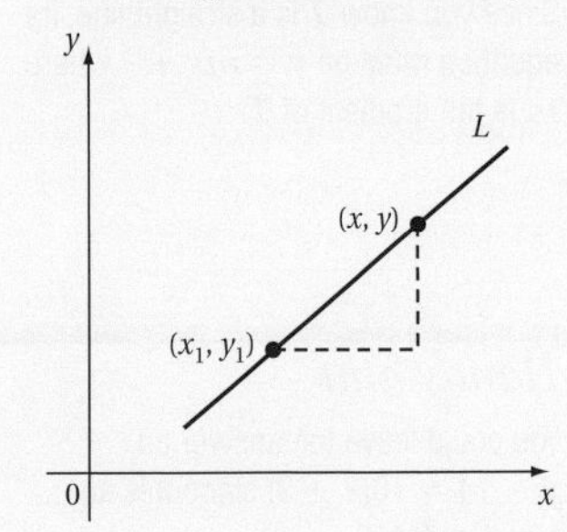

If (x,y) is any point on L, then $\text{grad}_L = \frac{y - y_1}{x - x_1}$

So $\frac{y - y_1}{x - x_1} = m$, leading to the result $y - y_1 = m(x - x_1)$.

Handy hint

This result is useful when m is a fraction but you want your equation to involve integers.

AS Level Example 6

Find the equation of the line which has gradient $-\frac{3}{2}$ and which passes through the point $A(4,5)$. Give your answer in the form $ay + bx = c$ where a, b and c are integers.

Working

You can use the result $y - y_1 = m(x - x_1)$, where the gradient $m = -\frac{3}{2}$ and the point $(x_1,y_1) = (4,5)$.

So an equation for the line is: $y - 5 = -\frac{3}{2}(x - 4)$

Multiply both sides by 2 to produce integers: $2y - 10 = -3(x - 4)$

Expand the bracket: $2y - 10 = -3x + 12$

Rearrange into the required form: $2y + 3x = 22$

So $2y + 3x = 22$ is an equation for this line.

> *Common error*
>
> $-3(x - 4)$ expanded is **not** $-3x - 12$.
>
> You must multiply all terms in the bracket by -3.

AS Level Example 7

The diagram shows the curve $y = x^2 + 6x - 2$ which passes through the point $P(2,14)$.

Find an equation for the tangent to this curve at P.

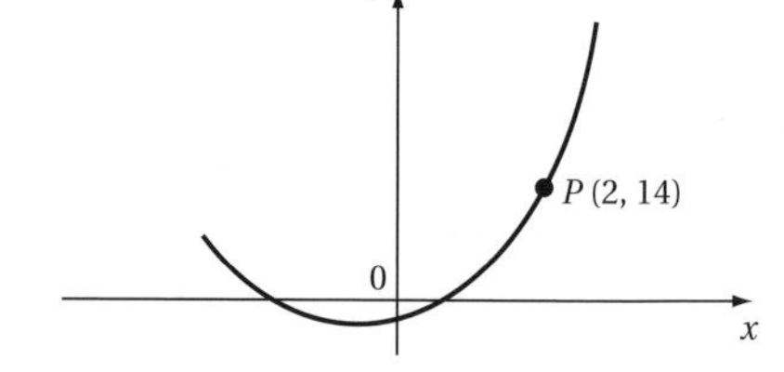

Working

You should sketch the tangent T whose equation you are trying to find.

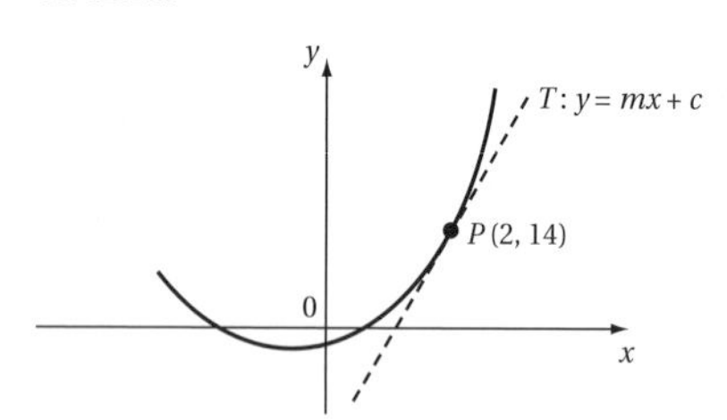

You first find m, the gradient of the tangent. To do this, you need to differentiate the curve equation.

Curve equation: $y = x^2 + 6x - 2$

Apply the rules: $\frac{dy}{dx} = 2x + 6$

Now substitute the x-coordinate of $P(2,14)$ into the gradient equation to find m.

Gradient equation: $\frac{dy}{dx} = 2x + 6$

Substitute in $x = 2$: $m = 2(2) + 6$

$= 10$

Now use the result $y - y_1 = m(x - x_1)$ to find an equation for T where $m = 10$ and $(x_1,y_1) = (2,14)$.

Equation of T: $y - y_1 = m(x - x_1)$

So: $y - 14 = 10(x - 2)$

Expand the bracket: $y - 14 = 10x - 20$

Add 14 to both sides: $y = 10x - 6$

So the tangent to this curve at point P has equation $y = 10x - 6$.

> *Handy hint*
>
> Since you know T is a straight line, its equation must be $y = mx + c$ where m is the gradient of T.

> *Handy hint*
>
> You could leave the answer as $y - 14 = 10(x - 2)$ since this an equation for T.

You can also find the equation of a **normal** to a curve at a given point.

Key point

The **normal** N to a curve at a point P is the line which

- passes through P **and**
- is perpendicular to the tangent T to the curve at P.

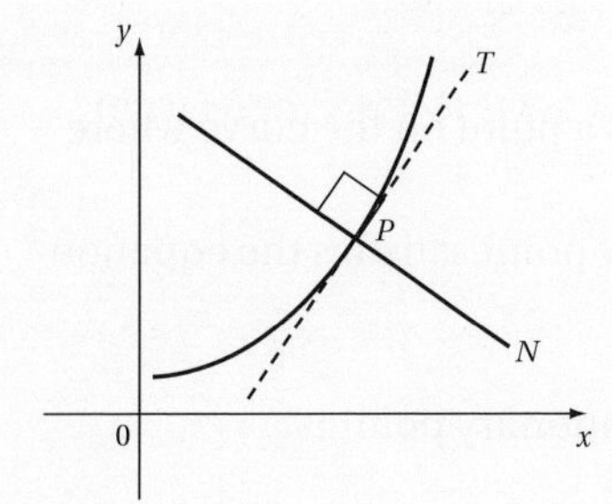

Handy hint

Perpendicular means 'at right angles to'. Refer to Section 3.4.

AS Level Example 8

The diagram shows the curve $y = 3x^2 - x^3$. The normal N to this curve at the point P where $x = 1$ is also shown.

Find an equation for N.

Give your answer in the form $ay + bx + c = 0$ for integers a, b and c.

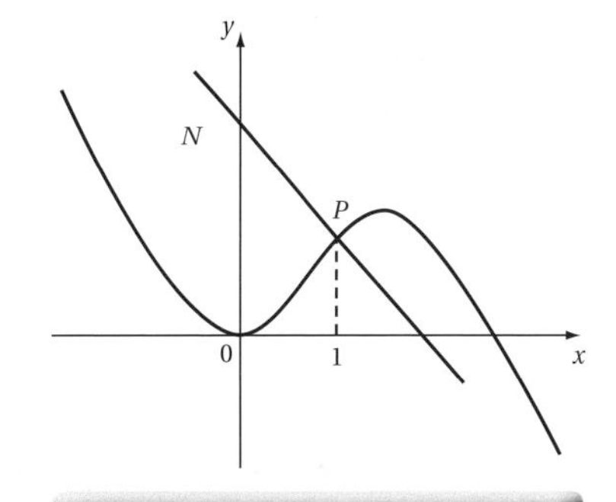

Working

You first need to find the y-coordinate of P.

The x-coordinate of P is 1 so substitute $x = 1$ into the curve equation

$$y = 3x^2 - x^3$$
$$= 3(1)^2 - (1)^3$$
$$= 2$$

So P has coordinates (1,2).

Handy hint

Find out any missing information about the curve before working out its gradient equation.

Now find the gradient equation $\frac{dy}{dx}$.

Curve equation: $y = 3x^2 - x^3$

Differentiate: $\frac{dy}{dx} = 6x - 3x^2$

Now find the gradient of the tangent and the normal to the curve at P.

The x-coordinate of P is 1 so substitute $x = 1$ into the gradient equation: $\frac{dy}{dx} = 6x - 3x^2$

So: $Grad_T = 6(1) - 3(1)^2$
$= 3$

Handy hint

There is no need to find the equation of T – you just need to know its gradient.

As N is perpendicular to T, the gradient of N is $-\frac{1}{3}$.

Handy hint

$Grad_T = \frac{3}{1} \xrightarrow[\text{change sign}]{\text{invert}} -\frac{1}{3} = Grad_N$.
Refer to Section 3.4.

Now use the result $y - y_1 = m(x - x_1)$ to find an equation for N where $m = -\frac{1}{3}$ and $(x_1, y_1) = (1,2)$.

Equation of N: $y - y_1 = m(x - x_1)$

So: $y - 2 = -\frac{1}{3}(x - 1)$

Handy hint

This is not an acceptable form for the answer because it includes a fraction.

Multiply both sides by 3 to produce integers:

$$3y - 6 = -(x - 1)$$

Expand the bracket: $3y - 6 = -x + 1$

Handy hint

Take care with the signs here: $-(x - 1) = -x + 1$

Rearrange the equation into the required form.

An equation for N is $3y + x - 7 = 0$.

In AS Maths you may be asked to locate the **stationary points** on a curve.

Key point

A **stationary point** on a curve is a point on the curve where the gradient is zero.

The x-coordinate of a stationary point satisfies the equation

$$\frac{dy}{dx} = 0$$

The tangent to the curve at a stationary point is horizontal.

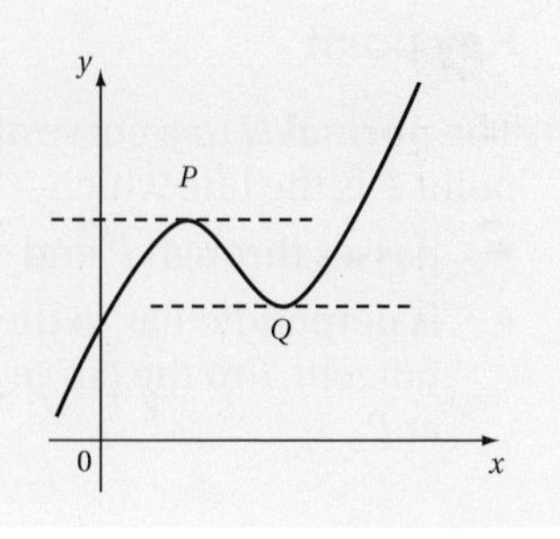

AS Level Example 9

The curve $y = x^2 - 6x + 13$ has a minimum stationary point P.

a Find the coordinates of point P.

b Hence, or otherwise, sketch the curve $y = x^2 - 6x + 13$.

Working

a The gradient of the curve is zero at a stationary point, so you need to solve the equation $\frac{dy}{dx} = 0$.

Curve equation: $y = x^2 - 6x + 13$

Differentiate: $\frac{dy}{dx} = 2x - 6$

Equation to solve: $\frac{dy}{dx} = 0$

So: $2x - 6 = 0$

Add 6 to both sides: $2x = 6$

Divide both sides by 2: $x = 3$

So the x-coordinate of P is 3.

You now find the y-coordinate of P by substituting $x = 3$ into the **curve equation**.

Curve equation: $y = x^2 - 6x + 13$

Substitute in $x = 3$: $= (3)^2 - 6(3) + 13$

$= 9 - 18 + 13$

$= 4$

So the stationary point P on this curve has coordinates (3,4).

Common error

The y-coordinate of P is **not** found by substituting $x = 3$ into the gradient equation $\frac{dy}{dx} = 2x - 6$.

b

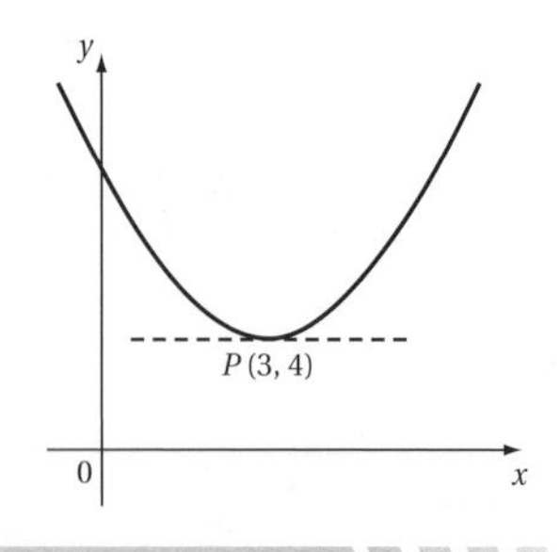

Handy hint

You could also sketch this curve by completing the square.

$y = x^2 - 6x + 13$

$= (x - 3)^2 - 9 + 13$

$= (x - 3)^2 + 4.$

This confirms the minimum point (vertex) has coordinates (3,4).

Taking it further

You will investigate stationary points in much more detail as you progress through your course. In particular, you will learn techniques for distinguishing between maximum and minimum points.

1 Practice: Numbers and indices

1.1 Fractions

1 Work out these. Give each answer in its simplest form.

a $\frac{2}{3} \times \frac{5}{6}$ **b** $\frac{3}{4} \div \frac{4}{5}$

c $\frac{3}{8} \times \left(\frac{2}{3}\right)^2$ **d** $\frac{6}{25} \div \frac{9}{10}$

2 Work out these. Give each answer in its simplest form.

a $\frac{1}{3} + \frac{2}{5}$ **b** $\frac{3}{5} - \frac{1}{10}$

c $\frac{4}{5} + \frac{1}{4} - \frac{1}{2}$ **d** $2 - \frac{1}{2} - \frac{3}{5}$

3 Convert these mixed numbers into top-heavy fractions.

a $2\frac{2}{3}$ **b** $4\frac{3}{5}$

c $5\frac{7}{11}$ **d** $\left(1\frac{2}{3}\right)^2$

4 Simplify these. Give each answer as a top-heavy fraction in its simplest form.

a $\frac{7}{\left(\frac{2}{5}\right)}$ **b** $\frac{\left(-\frac{8}{3}\right)}{2}$

c $\frac{\left(\frac{15}{4}\right)}{-3}$ **d** $\frac{\left(\frac{5}{6}\right)}{\left(\frac{4}{9}\right)}$

5 By converting mixed numbers to top-heavy fractions, find the value of these. Give each answer as a fraction in its simplest form.

a $\frac{6}{2\frac{1}{2}}$ **b** $\frac{8}{1\frac{1}{4}}$

c $\frac{\left(\frac{9}{2}\right)}{2\frac{2}{5}}$ **d** $\frac{3\frac{3}{5}}{6\frac{3}{4}}$

6 Use these formulae to find the value of y for the given value of x. Give each answer for y as a fraction in its simplest terms.

a $y = \frac{2}{3}x + 1, x = 4$

b $y = \frac{4x}{3} - \frac{3}{2}, x = \frac{5}{2}$

c $y = 3 - \frac{5x}{4}, x = -\frac{2}{3}$

d $y = \frac{1}{4}x^2, x = \frac{2}{3}$

e $y = \frac{4x^3}{5} + \frac{3}{4}, x = \frac{1}{2}$

f $y = \frac{6}{5x}, x = \frac{3}{2}$

7 A curve has equation $y = \frac{x}{x + 1}$. Find the y-coordinate of the point on this curve where

a $x = \frac{1}{2}$ **b** $x = -\frac{4}{3}$

8 The formula for converting the temperature C degrees Celsius into the temperature F degrees Fahrenheit is

$F = \frac{9C}{5} + 32.$

a What is 15 °C in degrees Fahrenheit?

b The minimum overnight temperature in a village was –7.5 °C. What was this temperature in degrees Fahrenheit?

9 The formula for converting the distance K kilometres into the distance M miles is

$M = \frac{5K}{8}$

a How far in miles is the distance $9\frac{1}{3}$ km?

b How far in km is the distance $3\frac{3}{4}$ miles?

10 Solve these equations. Give each answer as a fraction in its simplest form.

a $\frac{3}{4}x = 2$ **b** $-\frac{5}{3}x = \frac{1}{2}$

c $1\frac{4}{7}x = -1$ **d** $3\frac{1}{4}x = 2\frac{3}{5}$

11 A small rectangular garden lawn has length $4\frac{2}{5}$ metres and width w metres. The lawn has area 11 m².

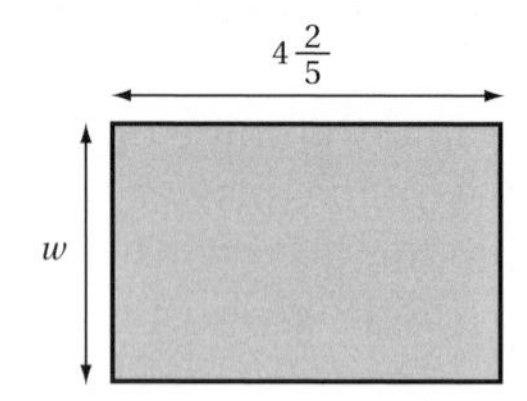

a Show that $w = \frac{5}{2}$.

b Find the perimeter of the lawn in metres. Give your answer as a mixed number.

1.2 Surds

1 Simplify these.

a $3\sqrt{5} + 4\sqrt{5}$ **b** $2\sqrt{3} \times 3\sqrt{2}$

c $\frac{4\sqrt{10}}{2\sqrt{5}}$ **d** $(2\sqrt{5})^2$

e $\left(-2\sqrt{2}\right)^3$ **f** $8\sqrt{\frac{7}{16}}$

2 Express these in simplified surd form.

a $\sqrt{45} + \sqrt{20}$ **b** $\sqrt{32} - \sqrt{18}$

c $2\sqrt{48} - 4\sqrt{27}$

3 By simplifying each surd, find the value of $\frac{\sqrt{50} + \sqrt{32}}{\sqrt{72} - \sqrt{18}}$.

4 Simplify these expressions.

a $\left(1 + \sqrt{2}\right)\left(2 + \sqrt{2}\right)$

b $\left(4 + \sqrt{3}\right)\left(1 - \sqrt{3}\right)$

c $\left(\sqrt{2} + 3\right)\left(2\sqrt{2} - 1\right)$

d $\left(2 + \sqrt{6}\right)^2$

e $\left(5 - 2\sqrt{3}\right)\left(4 + 3\sqrt{3}\right)$

f $\left(\sqrt{2} + \sqrt{3}\right)^2$

5 Express these fractions in the form $a + b\sqrt{3}$, where a and b are integers.

a $\frac{1}{2 + \sqrt{3}}$ **b** $\frac{12}{3 - \sqrt{3}}$

c $\frac{4\sqrt{3}}{\sqrt{3} + 1}$

6 Simplify these fractions.

a $\frac{5 + \sqrt{7}}{3 - \sqrt{7}}$ **b** $\frac{4 - \sqrt{3}}{2 - \sqrt{3}}$

c $\frac{3 + \sqrt{3}}{3 - 2\sqrt{3}}$

7 Show that $\frac{\sqrt{24} - 6}{3 - \sqrt{6}}$ is an integer, stating its value.

8 Given that $D = b^2 - 4ac$

a find the value of $\sqrt{D}$ when

i $a = 2, b = 4, c = 1$

ii $a = 3, b = -4, c = -2$

Give each answer in simplified surd form.

b Explain why $\sqrt{D}$ does not have a real value when $a = b = c$ where $b \neq 0$.

9 Express these in simplified surd form.

a $\sqrt{75}$ **b** $\sqrt{180}$

c $\sqrt{192}$ **d** $\sqrt{150} + \sqrt{96}$

10 ABC is a right-angled triangle. $AB = 4 + 2\sqrt{3}$, $AC = 4 - 2\sqrt{3}$.

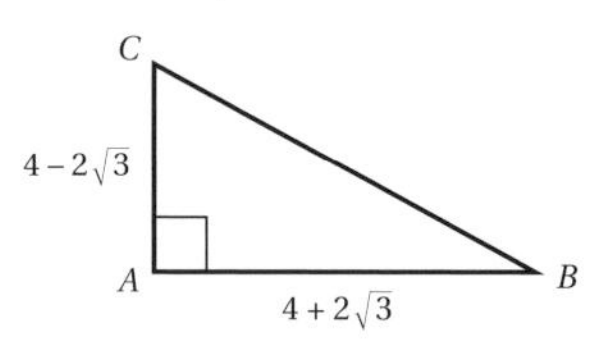

a Find the area of this triangle.

b Show that $BC^2 = 56$.

c Hence find the exact perimeter of this triangle. Give your answer in the form $a + b\sqrt{14}$, where a and b are integers to be stated.

11 a Find the greatest of these numbers. You may use a calculator if you wish.

$1 + \sqrt{3},\ 2 + 2\sqrt{3},\ 3 + \sqrt{3}$

b Show that these three numbers are sides of a right-angled triangle.

c Find the area of this triangle, giving your answer in the form $a + b\sqrt{3}$, where a and b are integers to be stated.

12 PQR is a right-angled triangle. $PQ = 5 + \sqrt{5}$, $PR = 3 + 3\sqrt{5}$.

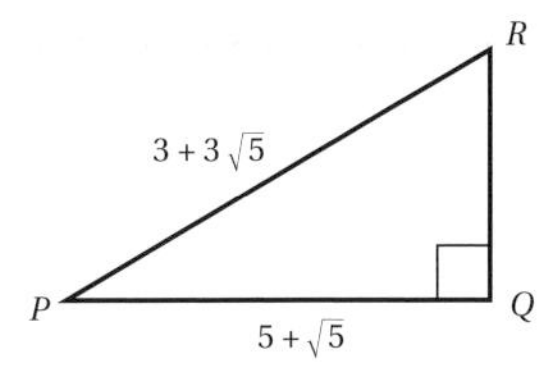

a Expand and then simplify $\left(1 + \sqrt{5}\right)^2$.

b Show that $QR^2 = 24 + 8\sqrt{5}$.

c Using your answer to part **a**, find the *exact* length QR.

d Show that the area and perimeter of this triangle are numerically equal.

1.3 Indices

1 Express each of these in the form 2^n where n is an integer.

a $2^3 \times 2^4$ **b** $(2^3)^3$

c 4^5 **d** $(2^4 \times 4^2)^3$

2 Express these as fractions in their simplest terms.

a 4^{-2} **b** 2^{-4}

c 5^{-3} **d** $2^{-1} \times 3^{-3}$

3 Evaluate these.

a $9^{\frac{1}{2}}$ **b** $4^{\frac{1}{2}} \times 27^{\frac{1}{3}}$

c $\dfrac{64^{\frac{1}{3}}}{16^{\frac{1}{4}}}$ **d** $32^{\frac{1}{5}} \times 17^0$

4 Evaluate these. Where appropriate, give answers as fractions in their simplest form.

a $4^{\frac{3}{2}}$ **b** $27^{\frac{2}{3}}$

c $8^{\frac{5}{3}}$ **d** $81^{\frac{3}{4}}$

e $16^{-\frac{1}{2}}$ **f** $8^{-\frac{1}{3}}$

g $125^{-\frac{2}{3}}$ **h** $64^{-\frac{4}{3}}$

5 By writing 16 as a power of 2, or otherwise, solve the equation $16^x = 32$.

6 Solve these equations.

a $8^x = 16$ **b** $16^x = 64$

c $9^x \times 3^x = 9$ **d** $\dfrac{8^x}{4^{x+1}} = 32$

7 Express these terms in the form ax^n where a is a real number.

a $\dfrac{4x}{2x^2}$ **b** $\dfrac{1}{2x^3}$

c $3x\sqrt{x}$ **d** $\dfrac{\sqrt[3]{x^2}}{4}$

e $\dfrac{2}{\sqrt{x}}$ **f** $\dfrac{3x}{\sqrt[3]{x}}$

g $\dfrac{3\sqrt{x^3}}{6x^2}$ **h** $\dfrac{10x}{\sqrt[4]{x^3}}$

8 Determine whether each of these statements is true or false. Use rules of indices to prove those which you think are true. For those statements that you think are false, give an example to show that it is incorrect.

a $a^n \times a^n = a^{2n}$ for all numbers a and positive integers n

b $a^n \times b^n = ab^n$ for all numbers a and b and positive integers n

c $a^{mn} = a^m \times a^n$ for all numbers a and positive integers m and n

d $a^n \times a^{-n} = 1$ for all non-zero numbers a and positive integers n

e $(a^n)^n = a^{2n}$ for all numbers a and positive integers n

9 Evaluate these expression using the given value of x. Give answers as top-heavy fractions where appropriate.

a $3x^{\frac{3}{2}}$ when $x = 4$

b $x^{\frac{1}{2}} - x^{-\frac{1}{2}}$ when $x = 9$

c $\frac{1}{2}x^{\frac{1}{3}} + 16x^{-2}$ when $x = 8$

d $12x^{-\frac{2}{3}} - 6x^{-\frac{1}{2}}$ when $x = 64$

e $4x^{\frac{1}{2}} + x^{-\frac{1}{2}}$ when $x = \frac{1}{4}$

f $4x^{-1} + \sqrt{2}\,x^{\frac{1}{2}}$ when $x = \frac{8}{9}$

10 a Express $\frac{3x^3 + 2}{x^2}$ in the form $ax + bx^n$, where a, b and n are constants.

b Express $\frac{2x^2 - 3x + 1}{2x^2}$ in the form $a + bx^{-1} + cx^{-2}$, where a, b and c are constants.

11 Express these as sums of powers of x.

a $\frac{(2x + 1)(x - 1)}{x}$

b $\frac{(3x + 2)^2}{x^3}$

c $\frac{x^2 + 3x - 6}{\sqrt{x}}$

d $\frac{(2 + \sqrt{x})^2}{x^2}$

12 A curve C has equation $y = \frac{(3x + 2)(2x + 3)}{x^2}$ where $x > 0$.

a Express y in the form $a + bx^{-1} + cx^{-2}$, where a, b and c are constants.

b Explain why, as x increases, the value of y approaches 6.

c Is there a point on this curve with y-coordinate 6?

2 Practice: Algebra 1

2.1 Basic algebra

1 Expand and then simplify these expressions.

a $2(a + 3) + 3(a - 1)$

b $3(b + 2) - 4(2b - 3)$

c $4(a + 2b) + 2(3a - 4b)$

d $a(2a + b) - b(a - 3b)$

2 Factorise fully these expressions.

a $2x^2y + xy^2$ **b** $10x^3y^2 - 4x^2y^3$

c $3x^4y^2z + 6x^3yz^2$ **d** $12x^4y^2 + 6x^2y^2 - 9xy$

3 Expand these expressions. Fully factorise answers where appropriate.

a $(2a^2b)^2$ **b** $(3ab^2)^3 + (3a^2b)^2$

c $(4a^2b^2)^2 - (2ab^3)^2$

4 Rearrange these equations to make the variable shown in square brackets the subject.

a $P = 3(Q + 4)$ [Q]

b $A = \frac{1}{2}(3B - 1)$ [B]

c $R + T = 3(T - 1)$ [T]

d $2(C - D) = 5(1 + 2D)$ [D]

e $U = \frac{1}{3}\sqrt{V + 2}$ [V]

f $M = \frac{\pi}{2}(N - 1)^3$ [N]

5 The volume V of a sphere with radius r is $V = \frac{4}{3}\pi r^3$.

a Re-arrange this formula to make r the subject.

b Find the radius of a sphere with volume 36π cm³.

6 The diagram shows a square $PQRS$ of side length $2x$ cm. A quarter circle, centre P and radius $2x$ cm, is inscribed inside the square.

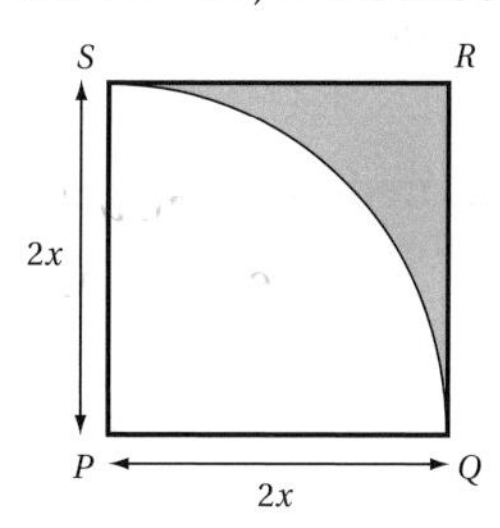

a Show that the area A of the shaded shape is given by the formula $A = 4x^2 - \pi x^2$.

b Make x the subject of this formula.

c Show that the perimeter of the shaded shape is given by the expression

$(4 + \pi)\sqrt{\frac{A}{4 - \pi}}$.

7 The diagram shows a right-angled triangle ABC, where $AB = m + 1$, $BC = m - 1$ and $AC = n$.

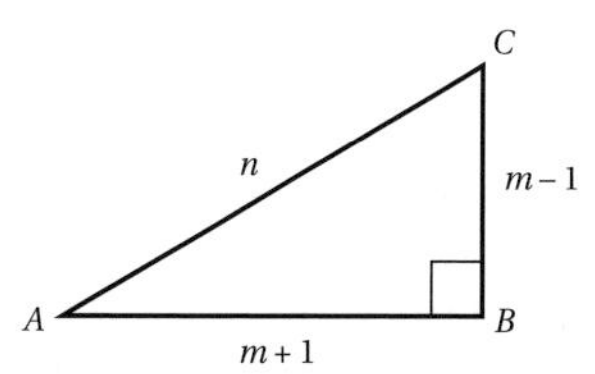

a If $t = \tan \hat{A}$ show that $m = \frac{1 + t}{1 - t}$.

b Find an expression for m in terms of n.

8 Rearrange these formulae to make x the subject.

a $y = (x + 3)^2$

b $y = 4(x - 1)^2 - 1$

c $y = \frac{(2x - 5)^2}{3}$

9 $P = \frac{2Q + 3}{Q}$

a Show that $P = 2 + \frac{3}{Q}$.

b Hence, or otherwise, make Q the subject of the formula $P = \frac{2Q + 3}{Q}$.

10 Make the letter indicated in square brackets the subject of these formulae.

a $A = \frac{B - 2}{B}$ [B]

b $C = \frac{D^2 + 4}{D^2}$ [D]

c $E = \frac{5 - 4F^3}{F^3}$ [F]

11 Make the letter indicated in square brackets the subject of these formulae.

a $A = \frac{B}{B - 2}$ [B]

b $C = \frac{D + 2}{2D + 3}$ [D]

c $E = \frac{F^2 + 3}{F^2 + 1}$ [F]

12 Simplify these fractions.

a $\frac{x^2 + 3x}{x}$ **b** $\frac{2x^4 + 4x^2}{x^2}$

c $\frac{3x^2 - 3x}{x - 1}$ **d** $\frac{x^2 - 2x^3}{2x - 1}$

13 Express these fractions in the required form. State the value of each constant.

a $\frac{3x^3 + 4x^2 + 6x}{3x}$ in the form $Ax^2 + Bx + C$ for constants A, B and C.

b $\frac{4x^3 - 3x}{2x^2}$ in the form $Ax - Bx^{-1}$ for constants A and B.

c $\frac{6x^4 + 9x^2 + 1}{3x^2}$ in the form $Ax^2 + Bx^{-2} + C$ for constants A, B and C.

d $\frac{12x^4 - 4x^3}{3x - 1}$ in for the form Ax^n for constants A and n.

2.2 Solving linear equations

1 Solve these equations.

a $3(a + 2) = 21$ **b** $8 + \frac{1}{2}(b - 2) = 11$

c $5(11 - 2c) = 25$ **d** $\frac{19 - 3d}{4} = 7$

2 Solve these equations.

a $4a - 3 = 2a + 7$ **b** $6b + 1 = 3 - 4b$

c $5(c - 2) = 8c + 2$

3 Solve these equations.

a $\frac{3x}{4} + 7 = 10$ **b** $\frac{x}{2} + \frac{x}{4} = 6$

c $\frac{x}{2} + \frac{2x}{3} = 7$

4 By expressing these as linear equations, find the value of x.

a $\frac{12}{x + 1} = 3$ **b** $\frac{8x}{2x + 3} = 1$

c $\frac{3}{x} + \frac{1}{2x} = 14$

5 For the equation $6x + 8y = 24$

a find the value of x when $y = 0$

b find the value of y when $x = 2$.

6 For the equation $5x - 2y = k$, where k is a constant, it is known that $y = 3$ when $x = 6$.

a Show that $k = 24$.

b Hence find the value of x when $y = 8$.

7 Solve these simultaneous equations.

a $3x + 2y = 12$
$2x + 3y = 13$

b $4x - 3y = 31$
$5x + 2y = 33$

c $8x - 6y = 13$
$3x - 5y = 9$

8 The diagram shows a rectangle $ABCD$. $AB = 2x + 1$, $AD = 3x - 2$, where all dimensions are in centimetres.

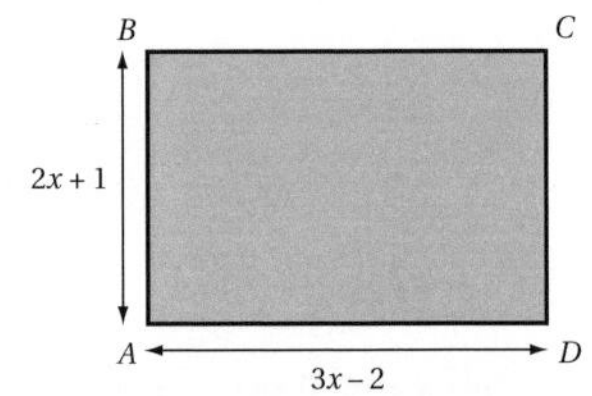

a Find an expression for the perimeter of this rectangle. Simplify your answer as far as possible.

The perimeter of this rectangle is 33 cm.

b Find the value of x.

c Hence find the area of this rectangle.

9 The shape in the diagram shows a rectangle $ABCD$ supporting a semi-circle, where $AB = (3x - 1)$ cm and the semi-circle has radius x cm. The centre of the semi-circle is the mid-point of BC.

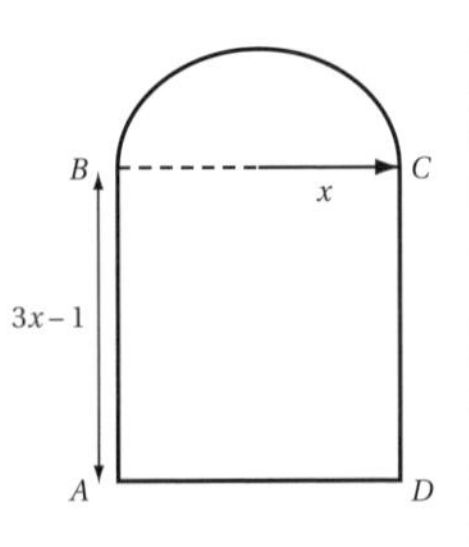

a Show that the perimeter P of the shape is given by the formula $P = (\pi + 8)x - 2$.

b Given that the perimeter of the shape is 555 cm, use a calculator to find the area of this shape. Give your answer to 3 significant figures.

10 By multiplying each of these equations by a suitable number, or otherwise, find the value of x for which:

a $\dfrac{x+3}{4} + \dfrac{x}{2} = 6$ **b** $\dfrac{2x-1}{3} - \dfrac{x+4}{6} = 5$

c $\dfrac{3x+1}{2} + \dfrac{x}{3} = 6$

2.3 Linear inequalities

1 Solve these inequalities.

a $5x + 2 \leqslant 12$ **b** $3x - 2 > -14$

c $3 - 4x \leqslant 11$ **d** $5x - 3 \geqslant 0$

2 Jane is solving the inequality $9 - 4x \leqslant -3$. Here is her working, which contains an error.

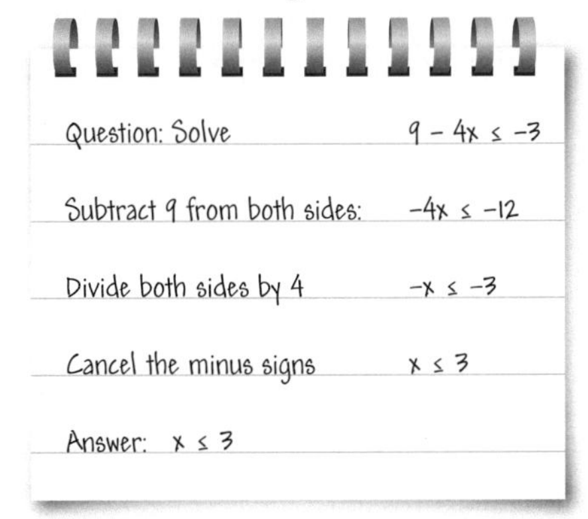

a Use the value of $x = 0$ to show that Jane's solution is incorrect.

b What error did Jane make?

c Give the correct solution to this inequality.

3 Solve these inequalities.

a $4x + 1 \geqslant x + 7$

b $2 - 5x < 11x - 6$

c $\frac{x}{4} - 4 > 2 - \frac{1}{2}x$

d $\dfrac{2x-1}{4} \leqslant 3x + 1$

e $5(1 - 3x) \geqslant 11 - 12x$

f $5 - \dfrac{4(x+2)}{3} < x$

4 Solve these inequalities.

a $2 \leqslant 3x - 1 \leqslant 11$

b $-3 < 2x + 3 \leqslant 1$

c $9 < 1 - 4x < 13$

5 **a** Solve the inequality $11 \leqslant 3x + 7 \leqslant 18$.

b Hence write down all the integer values of x which satisfy $11 \leqslant 3x + 7 \leqslant 18$.

6 Where possible, solve these inequalities.

a $3 \leqslant 3(x + 1) < 5$

b $12 < 3x < 6$

c $5 \leqslant 2 - 4x \leqslant 5$

7 By drawing a number line, or otherwise,

a find the set of values of x which satisfy **both** these inequalities: $4x - 1 \leqslant 15$ and $8 - 5x < -2$

b show that no value of x satisfies **both** these inequalities: $2x + 3 > 16$ and $4 - 3x \geqslant x - 20$.

8 Find the set of values of x which satisfy **neither** of these inequalities:

$\frac{1}{2}(x - 1) \leqslant 1$ and $5 - \frac{x}{4} < 4$.

9 Tom solves an inequality. He writes down the solution as $-4 < x < -7$.

a Explain why his answer does not make sense.

The inequality Tom was trying to solve is $23 < 11 - 3x < 32$.

b Solve this inequality.

10 Some wire is bent into the shape of the rectangle $ABCD$ (see diagram).

$AB = 5x - 4$ cm, $AD = 2x - 1$ cm.

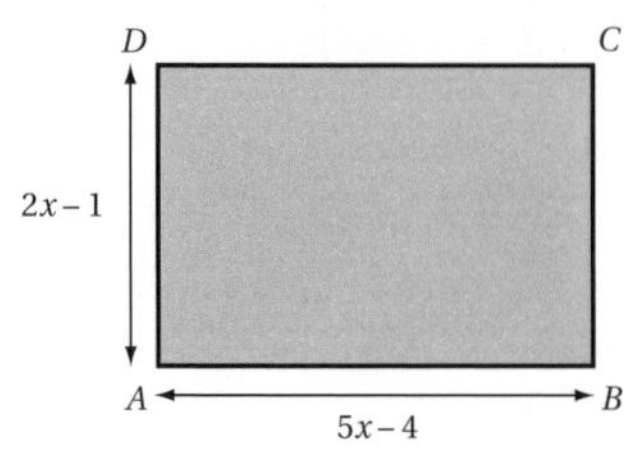

a Find the set of values of x for which the side AB is at least twice as long as side AD.

b Find an expression for the perimeter P of this rectangle.

The wire was originally cut from a piece with length 70 cm.

c What is the greatest possible *integer* value of x?

11 a Use algebra to solve the inequality $2x - 3 \geqslant 0$.

The diagram shows a drawing of the line with equation $y = 2x - 3$.

b Find the value of x where this line crosses the x-axis.

c How can your answer to **b** and the diagram be used to find the solution to the inequality $2x - 3 \geqslant 0$?

12 Use a **graphical** approach to solve these inequalities.

a $3 - 4x \geqslant 0$ **b** $3x + 2 > 0$

c $6x - 5 \leqslant 2x$

2.4 Forming expressions

Unless you are told otherwise, assume all lengths are in centimetres.

1 The diagram shows a rectangle $ABCD$. Point E is the mid-point of BC.

$AB = x$, $AD = 2x + 4$

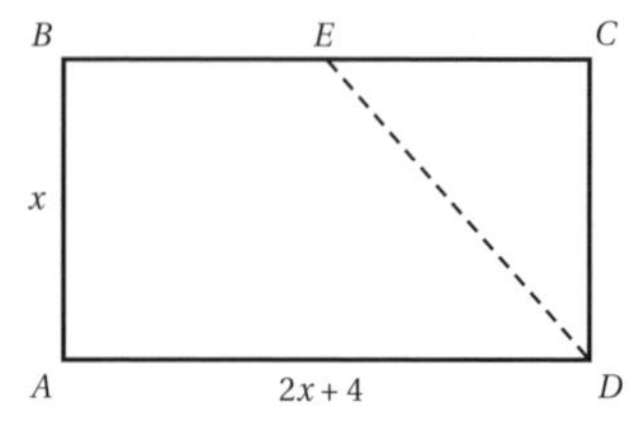

a Find an expression in terms of x for the perimeter of this rectangle.

b Show that the area of the trapezium $ABED$ is given by the formula

Area $= \frac{3}{2}x(x + 2)$.

2 The diagram shows the shape formed when a square of side length x is removed from the rectangle $ABCD$, where $AD = 2x + 5$ and $CD = 3x$.

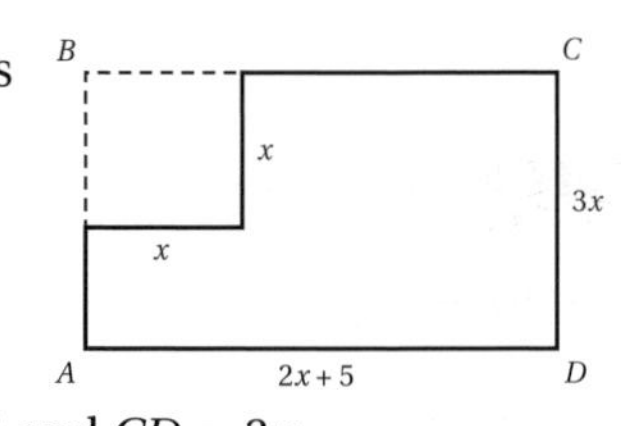

a Find, in factorised form, an expression for

i the perimeter of the shape

ii the area of the shape.

The area of the removed square is 49 cm^2.

b Find the area of the shape.

3 The diagram shows two circles with a common centre. The radius of the smaller circle is x cm. The (shortest) gap between the two circles is 3 cm.

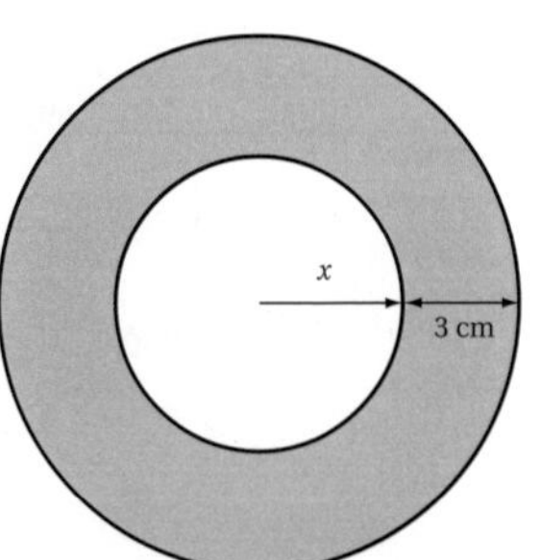

a Find an expression for the circumference of the larger circle. Leave π in your answer.

b Show that the area of the shaded region is given by the formula

Area $= 3\pi(2x + 3)$.

4 The diagram shows a circle with radius r and centre O. Points A and B on the circle are such that triangle AOB is right-angled.

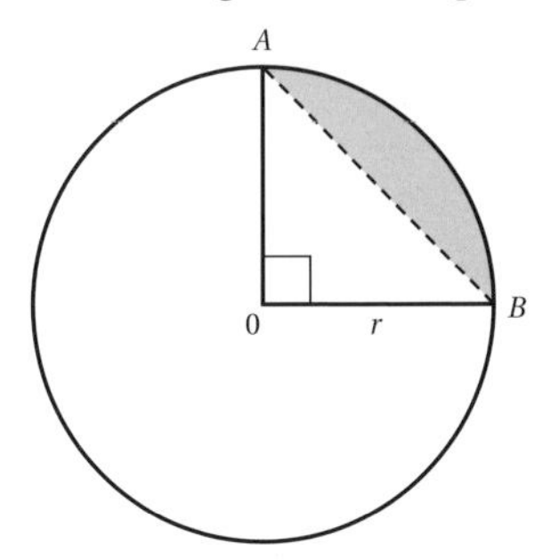

Handy hint

A sector of a circle looks like a slice of pizza!

a Show that the perimeter P of the sector OAB is given by the formula

$P = \frac{1}{2}r(4 + \pi)$.

b Find an expression in terms of r for the area between the line AB and the arc AB, as shaded in the diagram. Factorise your answer as far as possible.

5 The diagram shows a rectangle $ABCD$.

$AB = x + y$,
$BC = x - y$,
where $x > y$.

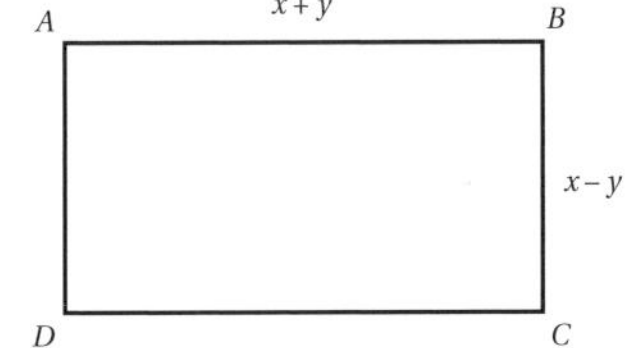

a Find an expression for the perimeter P of this rectangle.

The area of this rectangle is equal to the area of a square with side length y.

b Use this information to show that $x = ky$, stating the exact value of k.

6 An area of land is fenced off using some barbed wire and a wall. In the diagram the wire is represented by the edges AB, BC and CD. The side AD represents the wall, where $ABCD$ is a rectangle.

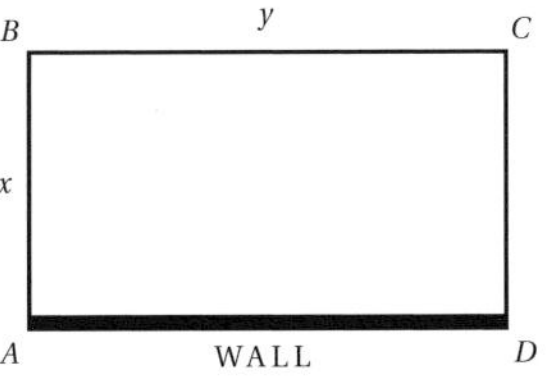

The total length of barbed wire used is 24 metres.

a Express this information as an equation involving x and y.

b Hence show that the area of this enclosure is given by the formula: Area $= 2x(12 - x)$.

c Find the enclosed area in the case when $ABCD$ is a square.

7 The diagram shows a cuboid with dimensions x, $2x$ and $3x$.

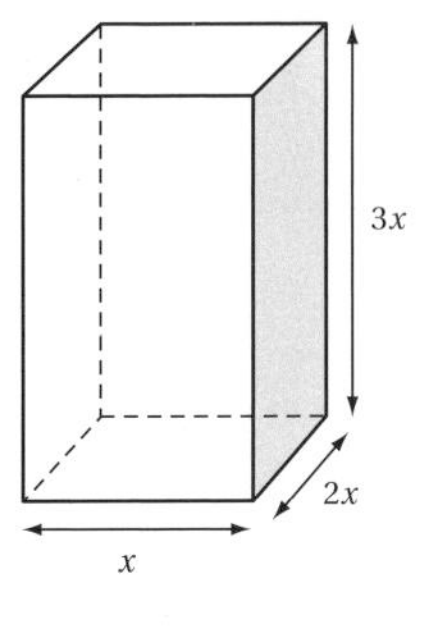

a Find an expression for the volume V of this cuboid. Simplify your answer as far as possible.

b Show that the surface area S of the cuboid is given by the formula $S = 22x^2$.

c Express the area of the side shaded in the diagram as a fraction of the surface area. Give your answer in its lowest terms.

The volume of this cuboid is 48 cm³.

d Find the surface area of this cuboid.

8 From a rectangle, four squares of side length x cm are cut from each corner.

Diagram 1 shows the net of the remaining shape.

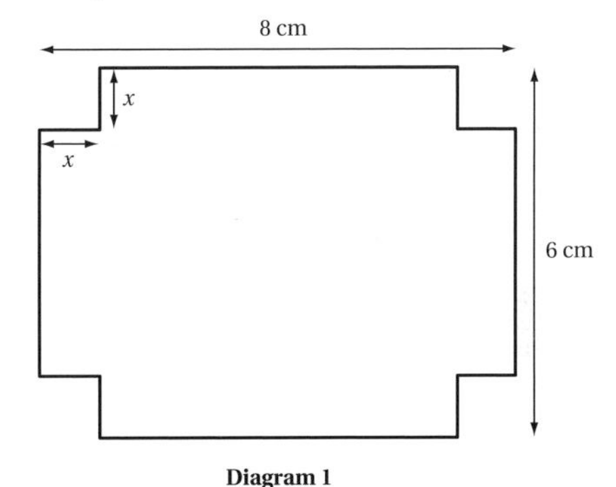

Diagram 1

a Find an expression in terms of x for the area of this net.

The sides of this net are folded at the corners to form a tray (see Diagram 2).

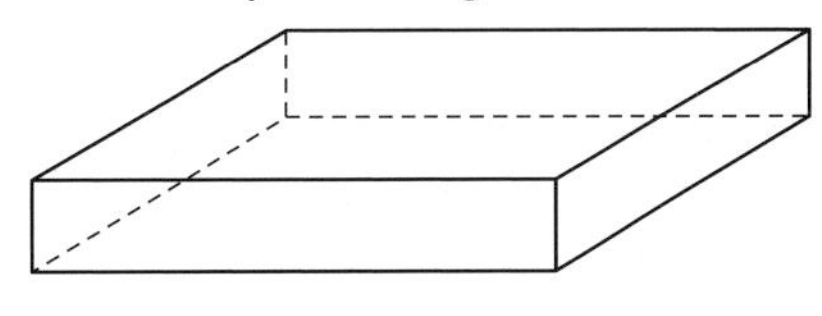

Diagram 2

b Show that the volume V of this tray is given by the formula $V = 4x(4 - x)(3 - x)$.

The surface area of this tray is 39 cm².

c Find the value of x.

d Hence calculate the volume of this tray.

9 The diagram shows a cuboid with base dimensions x cm by y cm. The cuboid has height 4 cm.

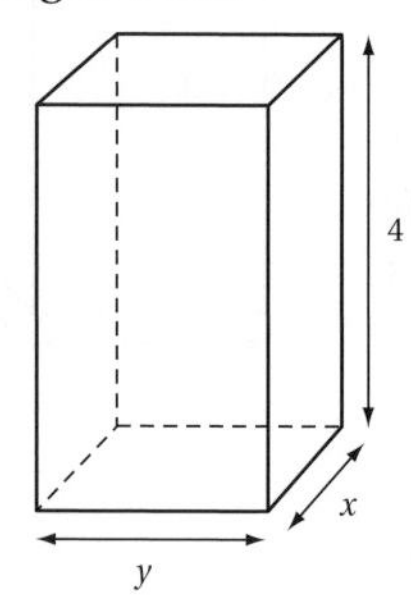

a Find an expression involving x and y for the surface area S of this cuboid.

The volume of this cuboid is 16 cm^3.

b Use this information to show that $xy = 4$.

c Find an expression for S in terms of x only.

10 The diagram shows a cylinder with radius r and height h.

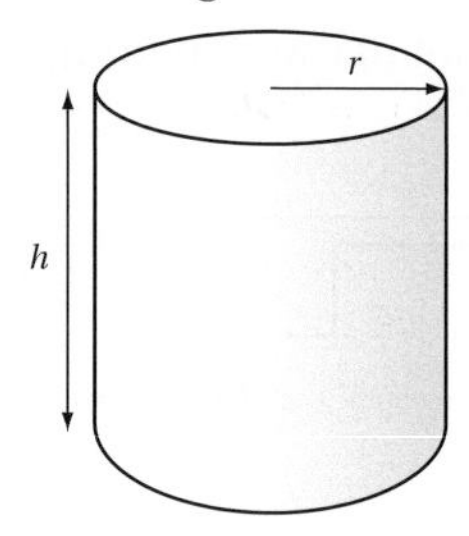

a Write down the volume V of this cylinder in terms of r and h.

It is given that the volume of this cylinder is 9π cm^3.

b Use this information to express r in terms of h. Simplify your answer as far as possible.

A straight metal rod, which is the longest that can be placed in the cylinder, has length L.

c Show that $L = \sqrt{\frac{36}{h} + h^2}$.

11 The rectangle $ABCD$ shown in Diagram 1 is curled so that the side AB meets the side CD to make the hollow cylinder with height 4 cm and radius r shown in Diagram 2.

$AB = 4$ cm and $BC = L$ cm.

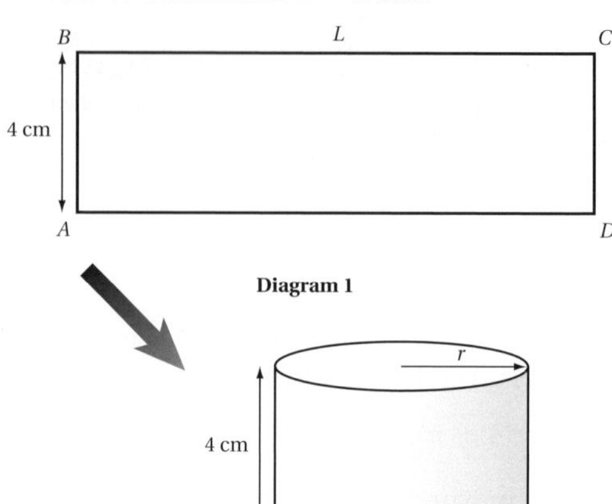

Diagram 1

Diagram 2

a Use Diagram 1 to find an expression in terms of L for the surface area of the cylinder.

b Explain why $r = \frac{L}{2\pi}$.

c Hence show that the volume V of this cylinder is given by the formula $V = \frac{L^2}{\pi}$.

A circular lid and a base each of radius r are added to the cylinder in Diagram 2.

d Show that the surface area S of this closed cylinder is given by the formula $S = \frac{L}{2\pi}(8\pi + L)$.

3 Practice: Coordinate geometry 1

3.1 Straight-line graphs

1 Find the gradient m and the y-intercept c of each of these lines.

a $y = 3x + 6$ **b** $y = 2 - 4x$

c $y = \frac{4x - 5}{2}$ **d** $y = -\frac{1}{3}(3 + 4x)$

Sketch, on separate diagrams, the lines with these equations. Label the points where the line crosses each axis with their coordinates.

2 By making y the subject of each equation, find the gradient m and the y-intercept c of these lines.

a $y - 2x + 1 = 0$ **b** $2y - 3x = 2$

c $4x - 3y = 1$ **d** $\frac{y}{4} + \frac{x}{2} = 3$

3 a Sketch, on the **same** diagram, the line $y - 2x = 1$ and the line $2y - 6x + 1 = 0$.

b Find the distance between the y-intercepts of these graphs.

4 a Sketch, on the **same** diagram, the line $y - 3x + 4 = 0$ and the line $3y + x = 6$.

b Find the distance between the x-intercepts of these graphs.

5 Express the equations of these lines in the form $ay + bx + c = 0$, where a, b and c are integers.

a $y = -\frac{1}{2}x - \frac{3}{2}$ **b** $y = \frac{1}{3} - \frac{2x}{3}$

c $y = -\frac{3}{4}x + \frac{1}{2}$ **d** $y = \frac{2}{3}x - \frac{5}{2}$

6 The diagram shows a sketch of the lines A, B, C and D. The lines have equations (1), (2), (3) and (4).

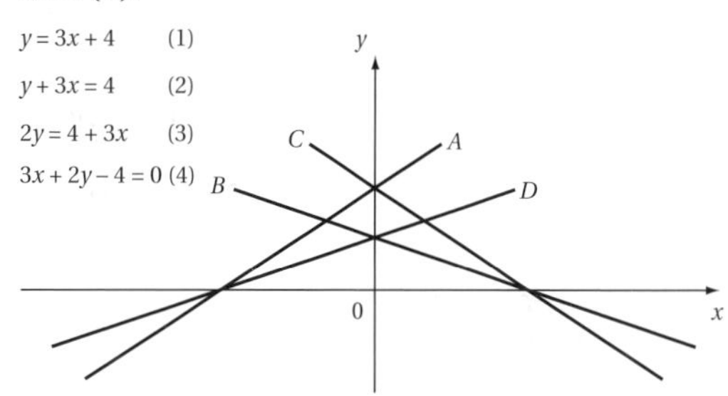

Match each line with its equation.

7 Determine which of these equations describe a straight line. For those that do, sketch their graphs on separate diagrams.

a $2(2x - y) = 1$

b $y(x + 1) = 4 + xy$

c $\sqrt{x^2 + y^2} = 4$

d $\frac{x}{y} = 2$

e $\frac{1}{x} + \frac{1}{y} = \frac{1}{2}$

f $3y(2x + 1) - 2x(3y + 2) = 6$

8 The line L has the same gradient as the line $4y + 2x + 3 = 0$.

a Find the gradient of L.

Given that this line L has the same y-intercept as the line $3y - 4x + 2 = 0$,

b find the y-intercept of L.

c Find the equation of L, giving your answer in the form $ay + bx + c = 0$ where a, b and c are integers.

9 The line L has equation $ay + bx = 10$, where a and b are constants.

The line crosses the y-axis at the point $(0,5)$ and crosses the x-axis at the point $(-2,0)$.

a Using this information, or otherwise, find the value of a and the value of b.

The point $P(4,q)$ lies on this line.

b Find the value of q.

3.2 The equation of a line

1 Find the gradient of the lines which pass through these points.

a $A(3,4)$ and $B(6,10)$

b $A(1,-3)$ and $B(3,7)$

c $A(-1,5)$ and $B(3,7)$

d $A(-4,-1)$ and $B(-6,5)$

2 The sketch shows the line L which passes through the points $A(0,3)$ and $B(2,2)$. This line crosses the x-axis at point C.

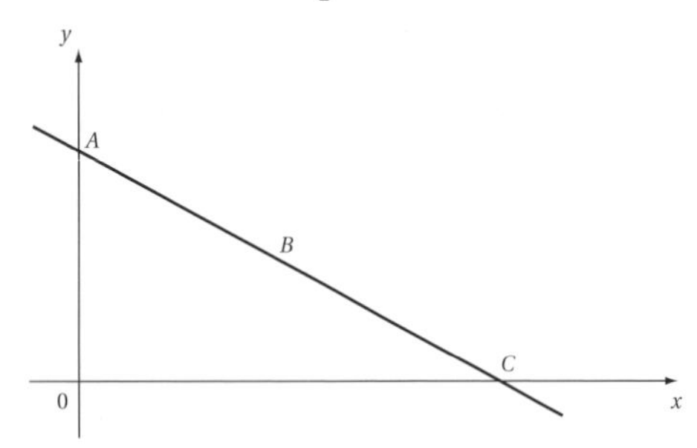

a Find the gradient of L.

b Find the equation of L. Give your answer in the form $ay + bx = c$ where a, b and c are integers.

c Find the coordinates of point C.

3 Find the equations of these lines.

a The line with gradient 4 which passes through the point $A(2,3)$.

b The horizontal line which passes through the point $C(35,-7)$.

c The line with gradient $-\frac{3}{2}$ which passes through the point $B(4,-2)$.

Give your answer for this line in the form $ay + bx = c$ where a, b and c are integers.

4 Find the equations of the lines passing through these points. Sketch each line on a separate diagram.

a $A(0,-1)$ and $B(5,14)$

b $A(2,5)$ and $B(4,1)$

c $A(-6,-4)$ and $B(10,8)$

d $A\left(\frac{1}{2},2\right)$ and $B(3,12)$

5 **a** A line passes through the points $A(1,5)$ and $B(3,p)$, where p is a constant. Given that the gradient of this line is 4, find the value of p.

b A line passes through the points $A(q,12)$ and $B(6,q)$, where q is a constant. Given that the gradient of this line is -7, find the value of q.

c A line passes through the points $E(r,r + 1)$ and $F(8,0)$, where r is a constant.

Given that the gradient of this line is $-\frac{1}{2}$, find the value of r.

6 The line L passes through the points $P(-4,-3)$ and $Q(4,9)$. This line crosses the y-axis at point A and the x-axis at point B.

Handy hint

Sketch the line L.

a Find an equation for L.

b Write down the coordinates of A.

c Find the coordinates of B.

d Find the area of triangle OAB, where O is the origin.

7 A line passes through the points $S(3,-2)$ and $T(12,-14)$. This line crosses the y-axis at point A and the x-axis at point B.

Handy hint

Sketch the line.

a Find the coordinates of A and the coordinates of B.

b Show that the distance $AB = \frac{5}{2}$.

8 A line has equation $y = mx - 3$ where m is a constant. The point $A(-5,7)$ lies on this line.

a Find the value of m.

b Determine whether or not the point $B(-7,10)$ lies on this line.

9 The line L has equation $2y - 4x + k = 0$ where k is a constant.

The point $A\left(\frac{5}{2},\frac{1}{2}\right)$ lies on L.

a Show that $k = 9$.

b Find the y-intercept of L.

c Find the area of the triangle formed by this line and the two coordinate axes.

10 The line L passes through the points $P(1,4k)$ and $Q(k,4)$, where k is a constant, $k > 1$.

a Show that the gradient of L is -4.

b Find the equation of L, giving your answer in terms of k.

Point R is where L crosses the x-axis.

c Show that the coordinates of R are $(k + 1,0)$.

d Find the value of k such that $PQ = QR$.

11 The line L_1 passes through the points $A(-3,-1)$ and $B\left(1,\frac{5}{3}\right)$.

a Show that the equation of L_1 can be written as $3y = 2x + 3$.

The line L_2 has equation $3x + 2y = 28$.

b Sketch, on a single diagram, the lines L_1 and L_2.

c Verify that point $P(6,5)$ lies on both L_1 and L_2.

Handy hint

For part **c** use the coordinates of P to confirm that P lies on each line.

d Find the area of the triangle formed by these lines and the y-axis.

3.3 Mid-points and distances

1 Find the coordinates of the mid-point of the line AB.

a $A(2,5)$, $B(10,3)$

b $A(5,-1)$, $B(-1,7)$

c $A(-6,11)$, $B(3,4)$

d $A\left(\frac{3}{2},\frac{5}{3}\right)$, $B\left(\frac{5}{2},\frac{1}{6}\right)$

2 For the points $A(3,2)$ and $B(p,q)$, where p and q are constants,

a find an expression for the coordinates of the mid-point of AB.

Given that this mid-point has coordinates $(4,1)$,

b find the value of p and the value of q.

3 Find, in terms of the constant k, the coordinates of the mid-point of AB. Simplify each answer as far as possible.

a $A(2,4)$, $B(k,2k)$

b $A(-2k,5)$, $B(0,2k+1)$

c $A(3k,2-k)$, $B(k,5k)$

4 The mid-point of AB is the point $C(2,3)$. If A has coordinates $(1,-2)$ find the coordinates of B.

5 Points $A(-4,4)$ and B are such that the mid-point of AB is the point $C(-3,7)$.

a Find the coordinates of B.

b Find the coordinates of the point D such that B is the mid-point of CD.

6 Points $A(p,3)$ and $B(14,q)$, where p and q are constants, are such that the mid-point of AB is the point $C(8,11)$. Point D is the mid-point of AC.

a Show that $p = 2$ and find the value of q.

b Find the coordinates of D.

It is given that the distance $AD = 5$.

c Find the distance DB.

7 The vertices of the square $ABCD$ have coordinates $A(1,3)$, $B(-3,7)$, $C(p,q)$ and $D(5,7)$, where p and q are constants.

a Find the coordinates of the mid-point of BD.

The diagonals of any square bisect each other.

b Using this information, or otherwise, find the coordinates of C.

8 Find the distance *AB* for these points. Give answers in simplified surd form where appropriate.

a $A(1,4)$, $B(6,2)$

b $A(2,-2)$, $B(5,7)$

c $A(-3,-1)$, $B(-5,9)$

d $A\left(-\frac{1}{2},\frac{3}{4}\right)$, $B\left(\frac{1}{2},\frac{3}{2}\right)$

9 Given the points $A(3,4)$, $B(6,-1)$ and $C(-2,7)$, prove that the triangle *ABC* is isosceles, but **not** equilateral.

10 A circle has diameter *AB*, where $A(4,-1)$ and $B(8,2)$.

a Find the coordinates of the centre of this circle.

b Show that the radius of this circle is $\frac{5}{2}$ units.

11 The point $C(2,-3)$ is the centre of a circle. The point $A(7,9)$ lies on this circle.

a Show that the radius of this circle is 13 units.

b Find the coordinates of the point *B* such that *AB* is a diameter of this circle.

12 The diagram shows the points $A(0,3)$, $B(3,9)$ and $C(5,8)$.

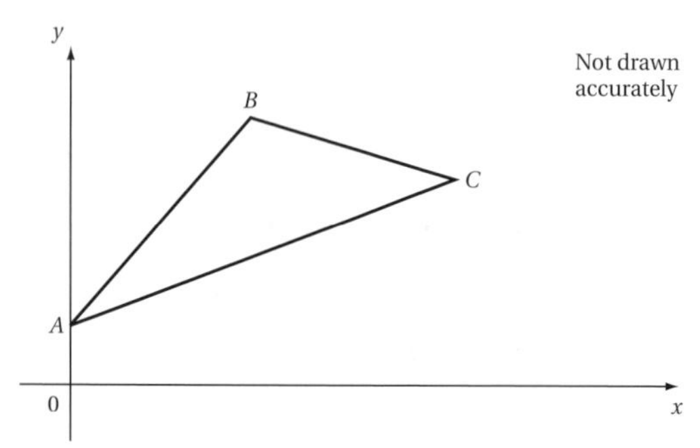

a Find the distances *AB*, *BC* and *AC*. Give each answer in surd form.

b Verify that $AB^2 + BC^2 = AC^2$.

c What information about the triangle *ABC* does the result of part **b** give?

Handy hint
Think Pythagoras!

13 The points $A(p,3)$ and $B(1,6)$, where p is a positive constant, are such that the distance $AB = p$.

a Show that $p = 5$.

b Find the coordinates of the mid-point *C* of *AB*.

c Find the circumference of the circle which has centre *C* and passes through point *A*. Leave π in your answer.

3.4 Parallel and perpendicular lines

1 Write down the gradient of any line which is

a parallel to the line $y = 5 - 3x$

b perpendicular to the line $y = 4x + 1$.

2 Find the gradient of any line which is

a parallel to the line $2y = 5x - 6$

b perpendicular to the line $3y + 4x = 1$.

3 By finding their gradients, show that these pairs of lines are parallel.

a $y = 2x - 4$
$y - 2x + 3 = 0$

b $2y - 3x = 1$
$y = \frac{4 + 3x}{2}$

c $2x + 4y - 3 = 0$
$2y + x + 1 = 0$

4 Show that these pairs of lines are perpendicular.

a $y = 3x + 4$
$3y + x = 3$

b $2y + 3x = 0$
$3y - 2x = 2$

c $y = \frac{11 - 5x}{3}$
$3x - 5y + 1 = 0$

5 The line *L* has equation $y = 2 - 4x$.

a Find the equation of the line which is parallel to *L* and which passes through the point $(0,3)$.

b Find the equation of the line which is perpendicular to *L* and which has the same y-intercept as *L*.

6 The line L has equation $4y - 3x = 11$.

a Find the gradient of L.

b Find the equation of the line which is perpendicular to L and which passes through the point $A(6,-6)$.

Give your answer in the form $ay + bx = c$ for integers a, b and c.

7 The line L has equation $y - 3x + 1 = 0$. The points $A(3,8)$ and $B(-1,k)$, where k is a constant, lie on L.

a Show that $k = -4$.

b Find the equation of the perpendicular bisector of AB. Give your answer in the form $ay + bx = c$, for integers a, b and c.

8 The diagram shows the triangle ABC where the vertices have coordinates $A(4,5)$, $B(8,10)$ and $C(13,6)$.

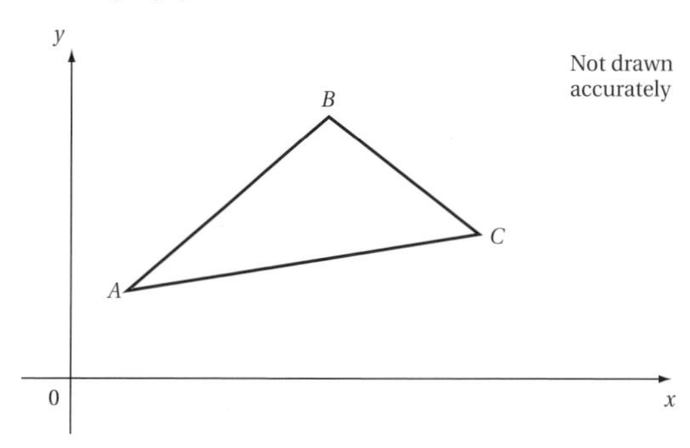

a Find the gradient of AB and the gradient of BC.

b Hence state the value of angle ABC.

c Show that triangle ABC is isosceles and hence state the value of angle BAC.

9 $A(2,7)$, $B(5,1)$ and $C(k,3)$, where k is a constant.

a Find the gradient of AB.

It is given that the line AB is perpendicular to the line BC.

b Show that $k = 9$.

c Find the equation of the perpendicular bisector of the line AC.

Give your answer in the form $ay + bx + c = 0$ for integers a, b and c.

10 The line L_1 passes through the points $A(3,2)$ and $B(9,0)$. The line L_2 is parallel to L_1 and passes through the point $C(7,4)$.

a Find the gradient of L_1.

b Hence find an equation for L_2.

c Show that the perpendicular bisector of AB passes through point C.

d Hence show that the perpendicular distance between L_1 and L_2 is $\sqrt{10}$ units.

3.5 Intersections of lines

1 i Sketch, on a single diagram, these pairs of lines.

ii Use algebra to find the coordinates of the point where the lines intersect. Give answers as top-heavy fractions where appropriate.

a $y = 4x - 9$
$y = 2x + 3$

b $y = \frac{x + 3}{2}$
$y = 2 - \frac{1}{2}x$

c $y = \frac{1}{3}x + \frac{3}{2}$
$y = \frac{2}{3} - \frac{1}{2}x$

2 Use algebra to find the coordinates of the points where these lines intersect.

a $2y + 3x - 5 = 0$
$y = 4 - 2x$

b $3y - 4x = 8$
$y = \frac{x + 2}{3}$

c $2y = x - 4$
$2x - 5y - 12 = 0$

3 Line L_1 has equation $8y + 6x = 5$. Line L_2 has equation $y = 2 - \frac{3}{4}x$.

a Use algebra to show that there is no solution to the simultaneous equations

$8y + 6x = 5$ and $y = 2 - \frac{3}{4}x$.

b What geometrical information does the result of part **a** give about the lines L_1 and L_2?

c Sketch, on the same diagram, the lines L_1 and L_2.

4 By expressing each equation in the form $y = mx + c$, for constants m and c, determine whether these pairs of lines intersect. For those that do, find the coordinates of the point of intersection.

a $4x + 3y = 6$
$3y - 4x + 18 = 0$

b $6x - 10y + 9 = 0$
$5y - 3x - 10 = 0$

c $6y + x = 2$
$3y = 1 + \frac{x}{2}$

5 **i** By solving these simultaneous equations, find the coordinates of the points where these lines intersect.

ii Sketch each pair of lines on a separate diagram.

a $2y - x = 5$
$2y + x = 9$

b $3y + x = 12$
$2y + 3x = 8$

c $2y - 3x + 9 = 0$
$5y - 2x = 5$

6 The diagram shows the lines L_1 and L_2. L_1 crosses the x-axis at the point A and has equation $y = 3x - 3$. L_2 crosses the x-axis at the point B and has equation $y = 7 - 2x$. These lines intersect at point C.

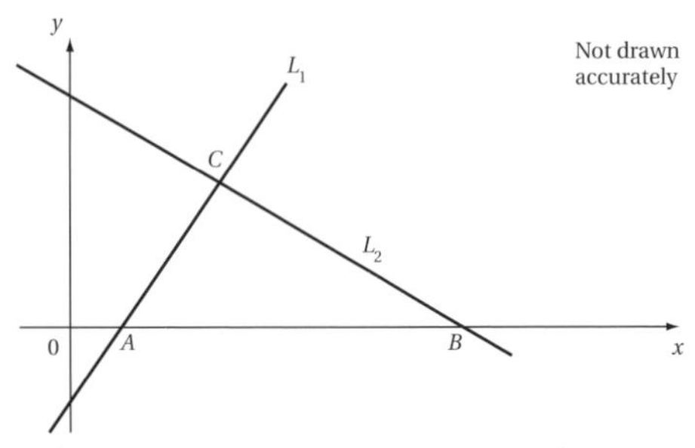

a Find the x-coordinate of point A and the x-coordinate of point B.

b Show that C has coordinates (2,3).

c Find the area of triangle ABC.

7 The diagram shows the line L_1 with equation $2y - x = 6$ and the line L_2 with equation $3y + 2x = 16$. These lines intersect at point C.

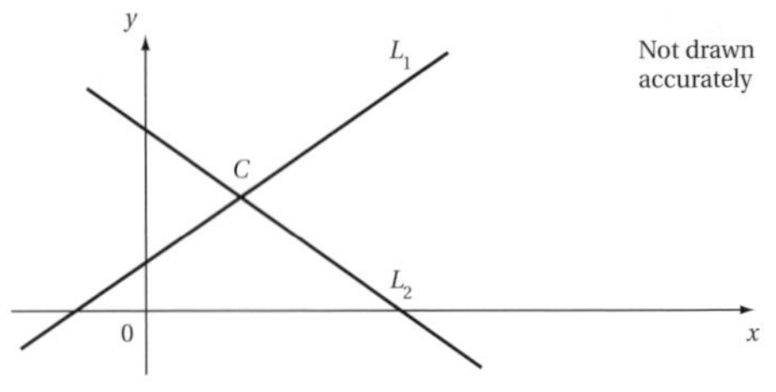

a Find the coordinates of point C.

b Find the area of the triangle formed by these lines and

i the y-axis **ii** the x-axis.

8 The line L_1 has equation $y = 6x - 3$. The line L_2 has equation $9y - 12x = 1$.

a Find the coordinates of the point A where these lines intersect.

b Find the equation of the line perpendicular to L_2 which passes through point A. Give your answer in the form $ay + bx = c$ for integers a, b and c.

9 The diagram shows triangle ABC formed by the intersection of three lines. Point A has coordinates (10,7). The line L_1 passes through A and C and has gradient $\frac{1}{4}$.

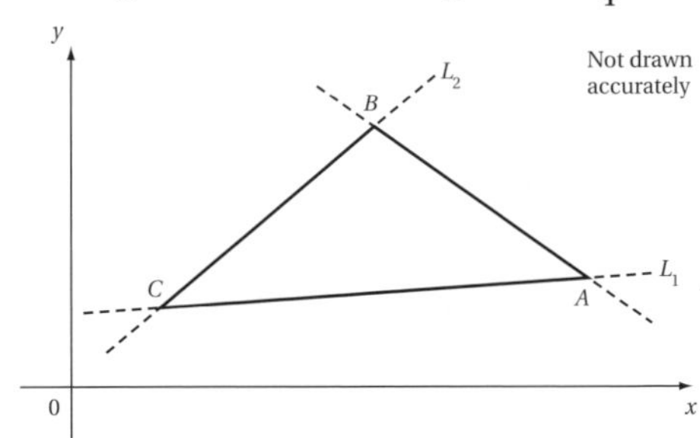

a Show that an equation for L_1 is $4y - x = 18$.

Line L_2 passes through the points B and C and has equation $2y - 3x = 4$.

b Find the coordinates of the point C.

The y-intercept of the line which passes through the points A and B is 17.

c Find the equation of this line and hence find the coordinates of point B.

d Find the distance BC, giving your answer in simplified surd form.

4 Practice: Algebra 2

4.1 Solving a quadratic equation by factorising

1 Use factorisation to solve these quadratic equations.

a $x^2 - 8x + 15 = 0$

b $x^2 + 5x - 14 = 0$

c $x^2 - 6x + 9 = 0$

d $x^2 - 9 = 0$

2 Solve these equations by using factorisation.

a $x^2 + 5x = 0$

b $x^2 = 6x$

c $2x^2 = 3x$

3 Simplify these equations and then solve them using factorisation.

a $2a^2 + 14a + 24 = 0$

b $3b^2 - 27b - 66 = 0$

c $4c^2 - 16c + 16 = 0$

4 Use factorising to solve these equations.

a $2x^2 - 9x - 5 = 0$

b $3x^2 + 2 = 5x$

c $2x^2 + 7x = 4$

d $x(x + 4) + 3x + 10 = 0$

e $x(2x - 3) = 2$

f $(x + 3)(x + 5) = 3$

5 Solve these equations by factorising.

a $p^2 + 10p - 56 = 0$

b $n^2 - 22n + 72 = 0$

c $2t^2 = 9(5 - t)$

6 The equation $x^2 - 15x + c = 0$, where c is a constant, is satisfied by the value $x = 2$.

a Show that $c = 26$.

b Hence find the other solution of this equation.

7 The equation $2x^2 + kx - 28 = 0$, where k is a constant, is satisfied by the value $x = -4$.

Find the other solution of this equation.

8 **a** By making the substitution $y = x^2$, express the equation $x^4 - 5x^2 + 4 = 0$ as a quadratic equation in y.

b Find the possible values of y and hence solve the equation $x^4 - 5x^2 + 4 = 0$.

9 Use factorisation to solve these equations. Use the hints if necessary.

a $x - 4\sqrt{x} + 3 = 0$ [Hint: let $y = \sqrt{x}$]

b $x^{\frac{2}{3}} - x^{\frac{1}{3}} - 6 = 0$ [Hint: let $y = x^{\frac{1}{3}}$]

c $2^{2x} - 9 \times 2^x + 8 = 0$ [Hint: let $y = 2^x$]

d $9^x - 10 \times 3^x + 9 = 0$

10 The diagram shows the trapezium $ABCD$, where the sides AB and CD are horizontal.

$AB = x$ cm, $CD = 2x$ cm and the height of the trapezium is $(x + 4)$ cm.

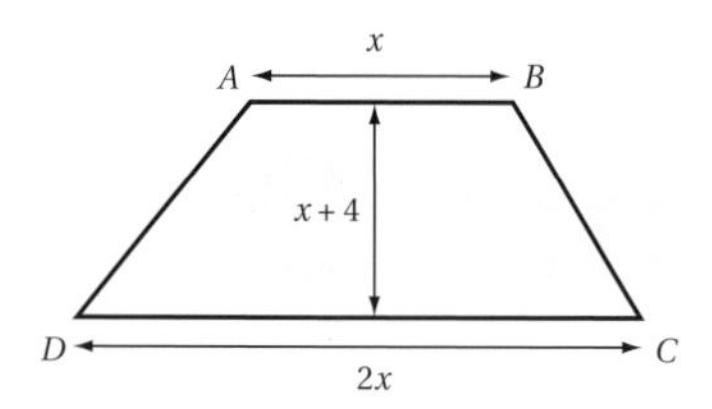

a Show that the area A of the trapezium is given by the formula

$A = \frac{3}{2}x^2 + 6x.$

The area of this trapezium is 48 cm^2.

b Show that $x^2 + 4x - 32 = 0$.

c Find the area of triangle ABC.

11 The diagram shows a right-angled triangle ABC.

$AB = x^2 - 1$ and $AC = 2x$, where all lengths are in centimetres.

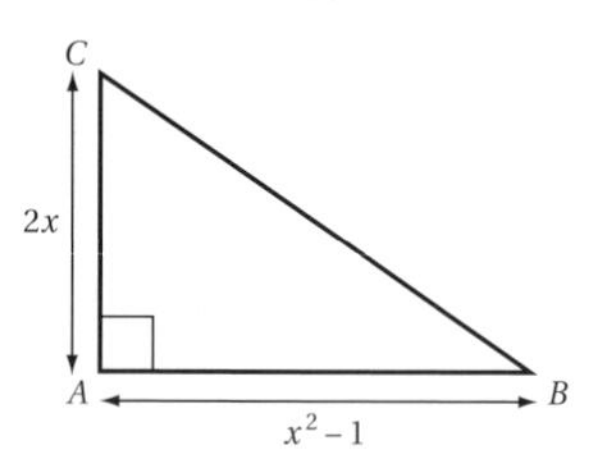

a Show that $BC = x^2 + 1$.

Given that the perimeter of this triangle is 112 cm,

b find the area of this triangle.

12 Consider the equation $(x - 1)(x + 2)(x + 3) = 0$.

You can solve this equation by making each bracket equal to 0.

So, if $(x - 1)(x + 2)(x + 3) = 0$ then either $x - 1 = 0$, $x + 2 = 0$ or $x + 3 = 0$.

The solutions to the equation $(x - 1)(x + 2)(x + 3) = 0$ are $x = 1$, $x = -2$ or $x = -3$.

$(x - 1)(x + 2)(x + 3) = 0$ is a **cubic** equation because if you were to expand all the brackets, the highest power of x in the expression would be 3.

Solve these cubic equations.

a $(x + 2)(x - 3)(x + 4) = 0$

b $(x - 4)(x + 1)(2x - 3) = 0$

c $x(x + 3)(x - 2) = 0$

d $(3x - 2)(x + 2)(5 - x) = 0$

e $(x + 1)(x^2 - 5x + 6) = 0$

f $(2x - 1)(x^2 - 9) = 0$

g $x^2(4x - 3) = 0$

h $x^3 + 3x^2 - 4x = 0$

Handy hint

In part **e**, factorise the quadratic.

4.2 Completing the square

1 Express these quadratics in completed square form.

a $x^2 + 6x + 10$ **b** $x^2 - 4x - 1$

c $x^2 + 8x + 16$

2 **a** Express $x^2 - 8x + 7$ in the form $(x - p)^2 - q$, where p and q are constants.

b Hence, or otherwise, solve the equation $x^2 - 8x + 7 = 0$.

3 Use completing the square to solve these equations. Give answers in simplified surd form where appropriate.

a $x^2 + 6x + 5 = 0$

b $x^2 - 2x - 2 = 0$

c $x^2 - 10x + 5 = 0$

d $x^2 - 3x + \frac{5}{4} = 0$

e $x^2 - x = \frac{1}{2}$

f $x^2 + 5x - \frac{1}{2} = 0$

4 **a** Express $3x^2 + 12x$ in the form $3(x + p)^2 - q$, stating the value of the constants p and q.

b Hence, or otherwise, express $3x^2 + 12x + 5$ in completed square form.

5 Express these quadratics in completed square form.

a $2x^2 + 12x + 17$ **b** $3x^2 - 18x + 31$

c $4x^2 - 4x + 3$

6 $f(x) = x^2 - 10x + 26$, where x is any real number.

a Find the minimum value of $f(x)$.

b State the value of x which gives this minimum.

7 $f(x) = 15 + 12x - x^2$, where x is any real number.

a Find the minimum value of the expression $x^2 - 12x - 15$.

b Hence write down the maximum value of $f(x)$ and the value of x at which this maximum occurs.

c Find the value of k for which the expression $k + 12x - x^2$ has a maximum value of 0.

8 a Express $x^2 - 2x + 3$ in the form $(x - p)^2 + q$, where p and q are constants.

b Hence show that the equation $x^2 - 2x + 3 = 0$ has no real solutions.

9 Some of the following equations have no real solutions. By completing the square, identify these equations. Where it appears, k is a non-zero constant.

a $x^2 + 2x + 6 = 0$

b $x^2 - 8x + 15 = 0$

c $x^2 + 10x + 25 = 0$

d $x^2 + 14x + 50 = 0$

e $x^2 + 4x + 4 + k^2 = 0$

f $x^2 + 2kx - k^2 = 0$

10 $f(x) = x^2 - 2x + 4$ where x is any real number.

a Use any appropriate method to solve the equation $f(x) = 3x$.

b Show that the equation $f(x) = 6x - 15$ has no real solutions.

11 a Express $x^2 - 8x + c$ in completed square form, where c is a constant.

The equation $x^2 - 8x + c = 0$ has no real solutions.

b Show that $c > 16$.

12 The equation $x^2 + 6x + k = 0$, where k is a positive integer, has at least one real solution.

Express as an inequality the possible values of k.

4.3 Solving a quadratic equation using the formula

1 Use the quadratic formula to find the exact solutions of these equations.

Simplify each answer as far as possible.

a $x^2 + 6x + 1 = 0$

b $x^2 + 4x - 3 = 0$

c $x^2 - 6x + 3 = 0$

d $2x^2 + x - 2 = 0$

e $3x^2 - 4x = 2$

f $2x^2 = 10x + 3$

2 Simplify these equations and then solve them using the quadratic formula.

Give answers in simplified surd form.

a $3x^2 - 2x = 2(x^2 + 3)$

b $4x^2 + 1 = (2 - x)^2$

c $2x(x - 3) = (x + 1)(x + 3)$

3 Apply the quadratic formula to show that these equations do **not** have any real solutions.

a $2x^2 + 5x + 4 = 0$

b $3x^2 - 7x + 5 = 0$

4 The diagram shows the point $A(2,4)$ and the point $B(p,p)$, where p is a constant, $p > 2$.

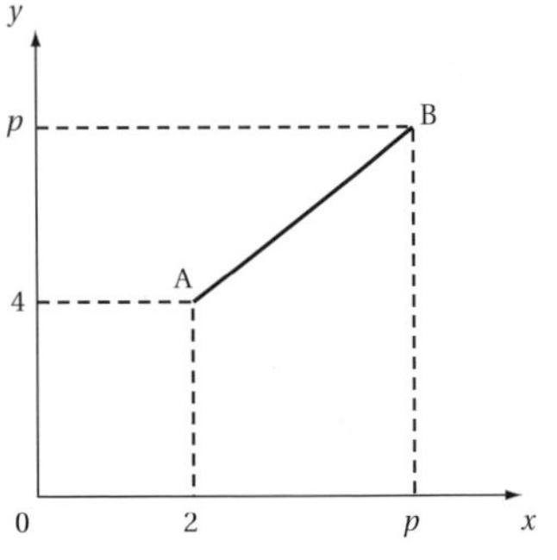

a Show that $AB^2 = 2p^2 - 12p + 20$.

Given that $AB = 4$,

b use the quadratic equation to find the value of p.

Give your answer in simplified surd form.

5 The diagram shows a shape consisting of a circle with radius r cm on top of a rectangle $ABCD$. The rectangle has height 3π cm and width equal to the diameter of the circle.

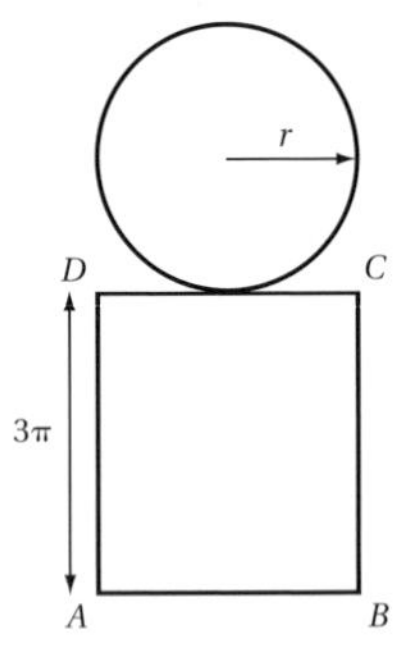

a Show that the area of the shape is given by $\pi r^2 + 6\pi r$.

Given that the area of the shape is 11π cm^2,

b show that r satisfies the equation $r^2 + 6r - 11 = 0$.

c Find the radius r. Give your answer in simplified surd form.

6 Find expressions, in terms of the constant k, for the solutions to these quadratic equations.

a $x^2 - 3x + k = 0$

b $x^2 - kx + 2 = 0$

c $kx^2 + 5x + 3 = 0$

7 $f(x) = x^2 - 2x + p$, where p is a constant.

a By applying the quadratic formula to the equation $f(x) = 0$, show that $x = 1 \pm \sqrt{1-p}$.

b Express as an inequality the values of p for which the equation $f(x) = 0$ has no real solutions.

8 $ax^2 - 4x + 3 = 0$ where a is a non-zero constant.

a For what range of values of a does this equation have no real solutions? Express your answer as an inequality.

b For which value of a does this equation have exactly one real solution? Find this solution.

9 $x^2 - 2kx - 1 = 0$, where k is any real number.

a Show that the solutions of this equation are given by $x = k \pm \sqrt{k^2+1}$.

b Briefly explain why this equation always has two distinct real solutions.

c Find, in terms of k, the sum of these two solutions

d Find the value of the product of these two solutions.

10 You may use a calculator in this question. The diagram shows a right-angled triangle ABC.

$AB = x$ cm, $BC = \dfrac{1}{x}$ cm and $AC = \sqrt{14}$ cm, where $x > 1$.

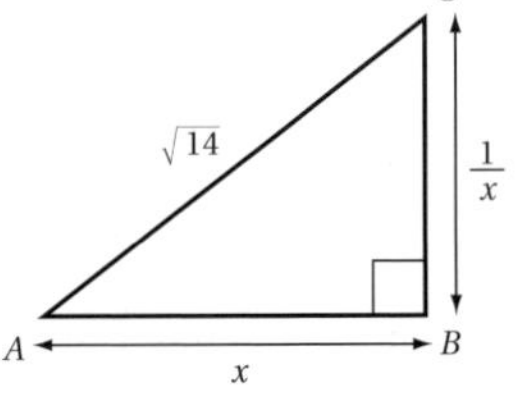

a Show that x satisfies the equation $x^4 - 14x^2 + 1 = 0$.

b Apply the quadratic formula to this equation to find the value of x^2. Give your answer in simplified surd form.

c Given that $\left(a + \sqrt{3}\right)^2 = 7 + 4\sqrt{3}$, find the value of the constant a.

d Hence write down the exact value of x.

e Show that the perimeter of this triangle is $4 + \sqrt{14}$ cm.

4.4 Solving linear and non-linear equations simultaneously

You may use a calculator for these questions.

1 Solve these simultaneous equations.

a $y = x + 3$
$x^2 + y^2 = 5$

b $y = 2x + 1$
$x^2 + y^2 = 10$

c $y = 3x + 2$
$2x^2 + y^2 = 3$

2 The diagram shows the circle $x^2 + y^2 = 5$ and the line $y = 3 - 2x$. Points A and B are where the line and the circle intersect.

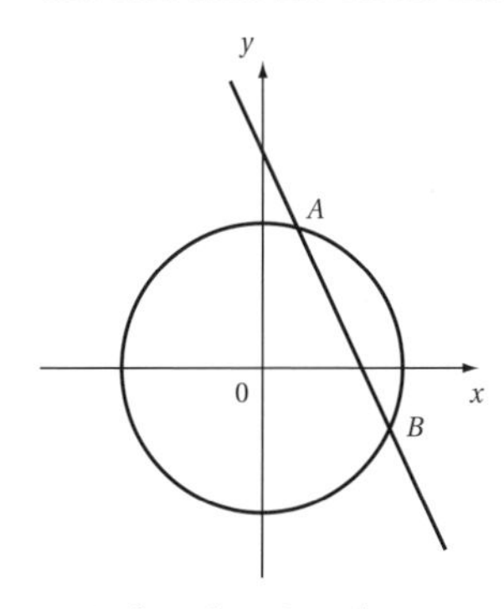

a Solve the simultaneous equations
$x^2 + y^2 = 5$
$y = 3 - 2x$
to find the coordinates of point A and the point B.

b Show that the distance AB is $\frac{8\sqrt{5}}{5}$ units.

3 Solve these simultaneous equations.

a $y - 2x + 1 = 0$
$y^2 - x^2 + 2x = 9$

b $x - 2y - 1 = 0$
$x^2 + 4y^2 = 5$

c $2x + 3y + 1 = 0$
$4x^2 - y^2 = 15$

4 The diagram shows the circle $x^2 + y^2 = 20$ and the line with equation $y = 2x + 10$.

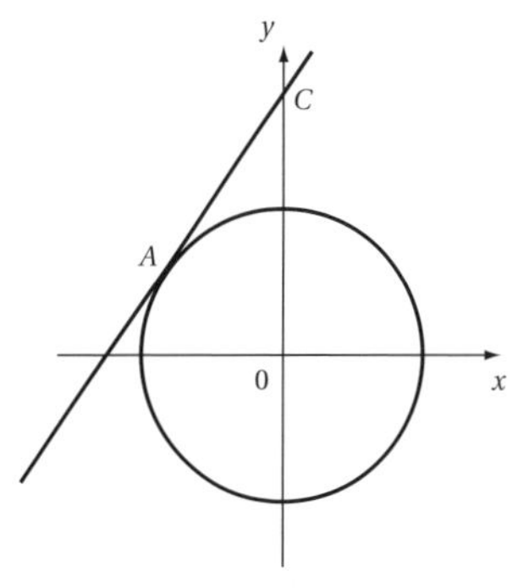

The line crosses the y-axis at point C and is a tangent to the circle at point A.

a Write down the coordinates of point C.

b Solve the simultaneous equations
$x^2 + y^2 = 20$
$y = 2x + 10$
to find the coordinates of point A.

c Find the area of triangle OAC.

5 A circle has equation $x^2 + y^2 = 10$ and a line has equation $y + 3x = 10$.

a Solve the simultaneous equations
$x^2 + y^2 = 10$
$y + 3x = 10$
to find the coordinates of any points where this line and circle intersect.

b What information about this line and circle does your answer to **a** give?

6 A circle has equation $x^2 - 2x + y^2 = 4$ and a line has equation $2y + x = 7$.

a Show that the simultaneous equations
$x^2 - 2x + y^2 = 4$
$2y + x = 7$
have no real solutions.

b What information about this line and circle does your answer to **a** give?

7 The shape in the diagram consists of two square metal plates welded together. The square *ABCG* has side length x metres. The smaller square *CDEF* has side length y metres.

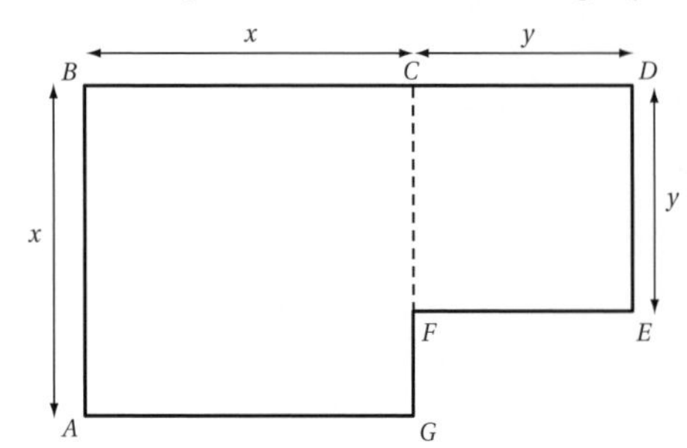

The area of the shape 5 m².

a Express this information as an equation involving x and y.

The perimeter of the shape is 10 m.

b Use this information to show that $2x + y = 5$.

c Solve a pair of simultaneous equations to find the dimensions of each plate.

8 The diagram shows a right-angled triangle *ABC*. $AB = x$ cm, $AC = y$ and $BC = 6$ cm.

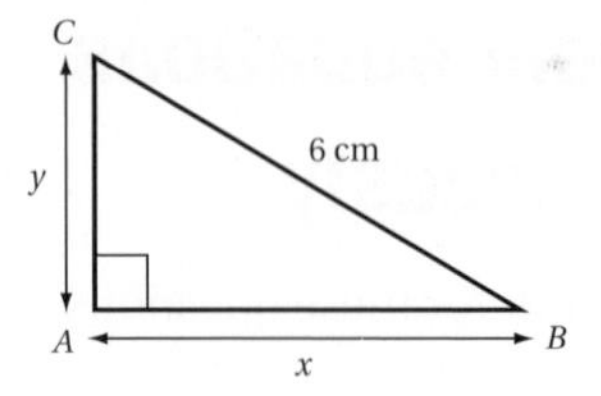

a Use the given information to write down an equation connecting x and y.

The triangle has perimeter 14 cm. It can be assumed that $x > y$.

b Solve a pair of simultaneous equations to find the exact length of side *AB* and the exact length of side *AC*.

c Find the area of triangle *ABC*.

5 Practice: Coordinate geometry 2

5.1 Transformations of graphs

1 The diagram shows the graph of $y = f(x)$ which crosses the y-axis at the point (0,7). Point $P(-2,3)$ is the minimum point on this graph.

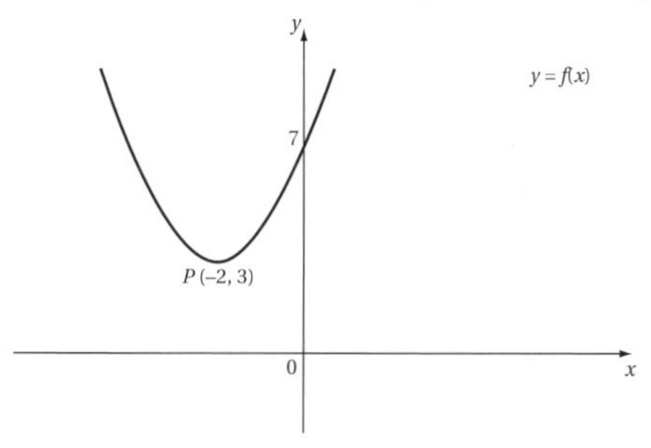

Sketch, on separate diagrams, the graphs of

a $y = f(x) + 2$ **b** $y = 2f(x)$

c $y = f(2x)$ **d** $y = f(x - 2)$.

Indicate on each sketch the coordinates of the minimum point and the y-intercept of each graph

2 The diagram shows the graphs F and G. The graph G is a translation of graph F by the vector $\begin{pmatrix} 3 \\ -4 \end{pmatrix}$.

Refer to Section 5.1 if you are not sure what this means.

Under this translation, point $P(1,4)$ on F is mapped to point P' on G, and point Q on F is mapped to point $Q'(6,-2)$ on G.

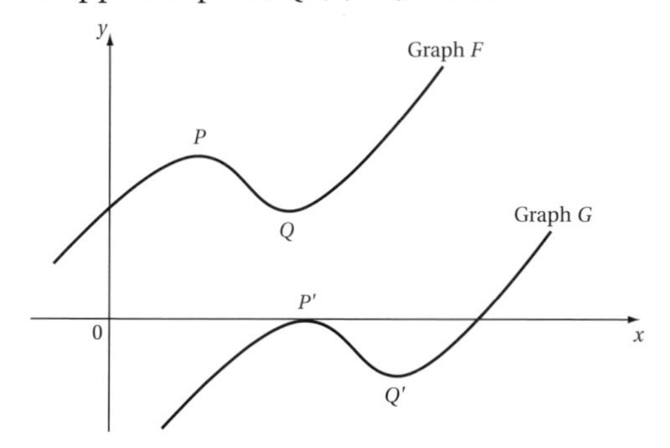

a Find the coordinates of

i point P' **ii** point Q.

b Sketch the graph of $y = \frac{3}{2}f(x)$.

Label the images of P and Q with their coordinates.

3 The diagram shows the graph of $y = f(x)$ and the graph of $y = f(ax)$, where a is a positive constant. The graph $y = f(x)$ crosses the x-axis at the points where $x = -1$ and $x = 6$. Both graphs cross the y-axis at the point (0,2).

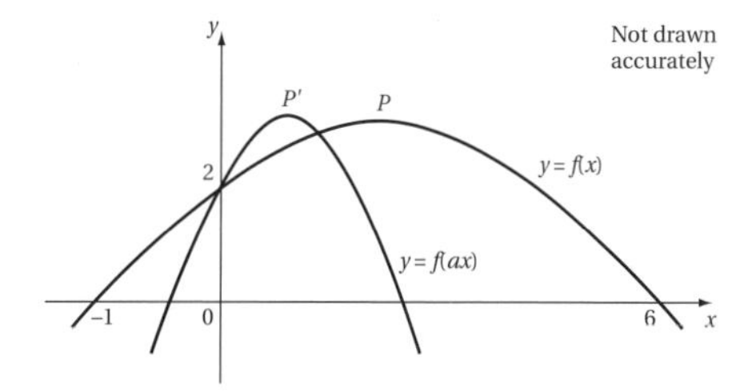

a Describe, in terms of a, the transformation which maps the graph of $y = f(x)$ onto the graph of $y = f(ax)$.

Under this transformation, the point $P(2,3)$ is mapped to the point $P'\left(\frac{1}{2},3\right)$.

b Find the value of a.

c For this value of a, find the coordinates of the points where the graph of $y = f(ax)$ crosses the x-axis.

d Sketch, on separate diagrams, the graphs of

i $y = f(x + 2)$ **ii** $y = 4f(4x)$.

On each sketch, mark the axis-crossing points with their values and the maximum point with its coordinates.

4 The diagram shows the graph of $y = f(x)$. The graph crosses the y-axis at the point (0,3) and the x-axis at the point (5,0). The maximum point P on this graph has coordinates (2,5).

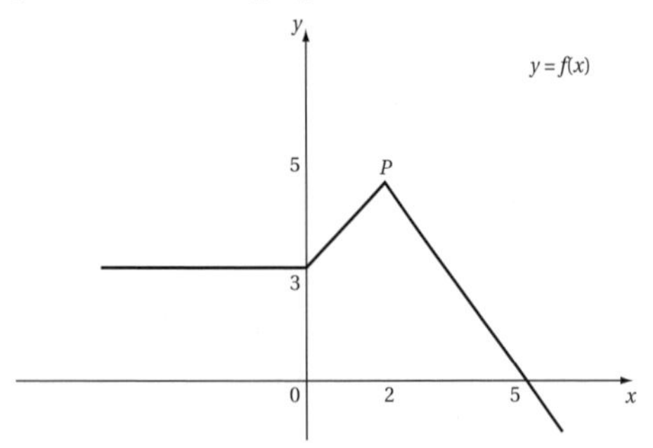

a Sketch, on separate diagrams, the graphs of
i $y = -f(x)$ **ii** $y = f(-x)$.

Indicate on each sketch the coordinates of the axis-crossing points and the maximum or minimum point of each graph.

b Sketch, on a new diagram, the graph of $y = -f(-x)$. Indicate the coordinates of the axis-crossing points and minimum point of this graph.

c Describe in as much detail as you can the **single** transformation which maps the graph of $y = f(x)$ to the graph of $y = -f(-x)$.

5 The diagram shows the complete graph of $y = f(x)$. The graph crosses the y-axis at the point (0,5) and the x-axis at the points (2,0) and (6,0). Point $P(4,-2)$ is the minimum point on this graph.

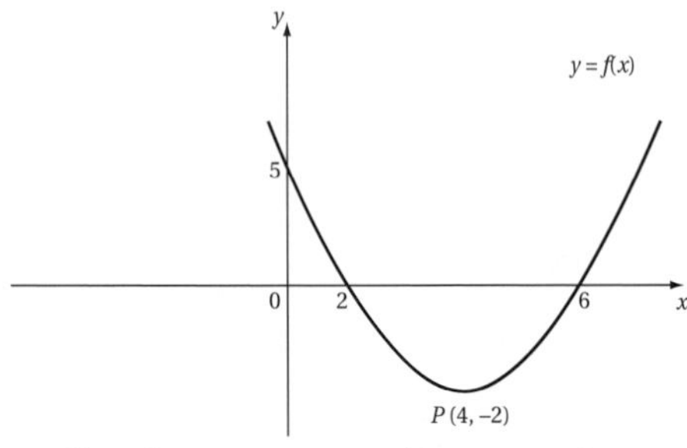

a Sketch, on separate diagrams, the graphs of
i $y = f(x) + 3$ **ii** $y = f\left(\frac{1}{2}x\right)$
iii $y = -f(2x)$ **iv** $y = 2f(3x)$.

Indicate on each sketch the coordinates of any axis-crossing points and the minimum or maximum point of each graph.

Under the translation $\begin{pmatrix} -2 \\ k \end{pmatrix}$, where $k > 0$, the graph of $y = f(x)$ is mapped to a graph G which crosses the x-axis at a single point.

b State the value of k.

c Find the coordinates of the points where G crosses each axis.

5.2 Sketching curves

1 For each of these,
i sketch the graph of each equation
ii label the vertex of each graph with its coordinates
iii describe the single translation which maps the graph of $y = x^2$ to the sketched graph.

a $y = x^2 + 4$

b $y = (x - 4)^2$

c $y = (x + 3)^2 + 2$

2 a Express $x^2 + 8x + 12$ in completed square form.

b Hence sketch the graph of $y = x^2 + 8x + 12$. Label the y-intercept with its value.

c Write down the equation of the line of symmetry for this graph.

3 Sketch these curves. On each sketch label the y-intercept with its value and the vertex with its coordinates.

a $y = x^2 + 2x + 5$

b $y = x^2 - 6x + 7$

c $y = x^2 + 3x - 1$

4 a State the sequence of transformations which map the graph of $y = x^2$ to the graph of $y = -x^2 + 1$.

b Sketch the graph of $y = -x^2 + 1$.

5 Sketch, on separate diagrams, these curves. On each sketch label the y-intercept with its value and label the vertex with its coordinates.

a $y = -(x - 2)^2$

b $y = 2 - (x + 3)^2$

c $y = 3 + 4x - x^2$

6 The diagram shows the graph of $y = x^2 + bx + c$, where b and c are constants.

The vertex of this graph has coordinates $(3, -4)$.

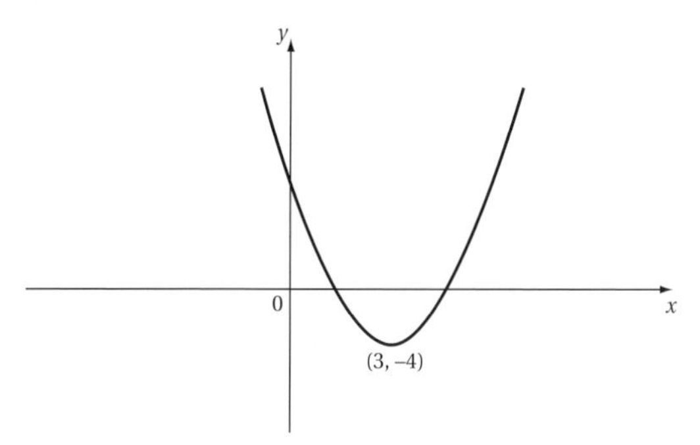

a By considering this graph as a transformation of $y = x^2$, express the equation of this graph in the form $y = (x - p)^2 - q$, stating the value of each of the constants p and q.

b Hence show that $b = -6$.

c Write down the value of the y-intercept of this graph.

7 The diagram shows the graph of $y = (x + p)^2 + q$, where p and q are constants.

The curve passes through the x-axis at points $A(1,0)$ and $B(7,0)$, and is symmetrical in the dotted vertical line.

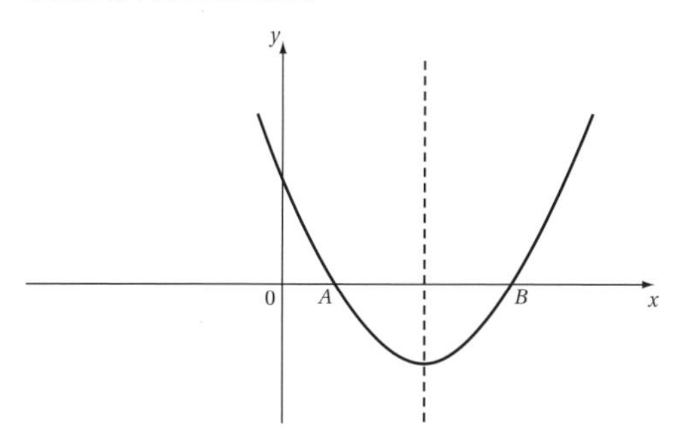

a Write down the equation of the line of symmetry shown on the diagram.

b Write down the value of p.

c Find the value of q.

d Describe the single transformation which maps the graph of $y = x^2$ to the graph in the diagram.

e Find the coordinates of the y-intercept of the graph in the diagram.

8 On a single diagram, sketch the graph of $y = x^2$ and the graphs of these equations.

Handy hint

See Section 5.1 for help with sketching $y = af(x)$.

a $y = 2x^2$

b $y = -3x^2$

c $y = 4x^2 - 1$ **d** $y = 2(x + 1)^2$

e $y = -4(x - 2)^2$ **f** $y = \frac{1}{2}(x + 3)^2 + 4$

9 The diagram shows the graph of $y = \sin x$ for $0° \leqslant x \leqslant 360°$.

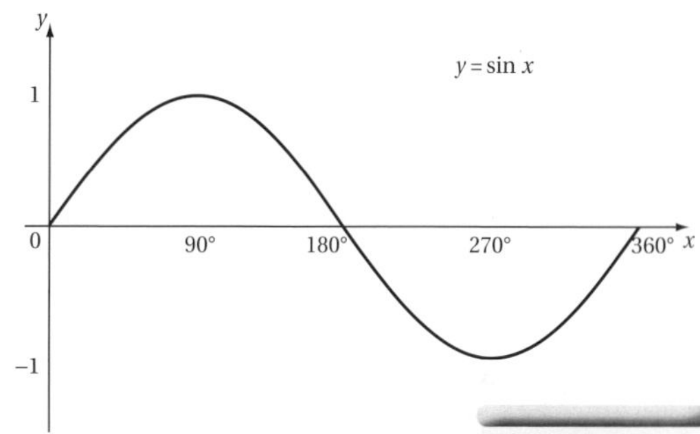

Sketch, on separate diagrams, these graphs for $0° \leqslant x \leqslant 360°$.

Handy hint

See Section 5.1 for help with sketching $y = f(ax)$.

a $y = 1 + \sin x$

b $y = 3 \sin x$ **c** $y = \sin 2x$

Indicate on each sketch the coordinates of any maximum or minimum points.

10 The diagram shows one complete cycle of the graph of $y = \cos x°$, where $x \geqslant 0$.

Point P is a minimum point on this graph.

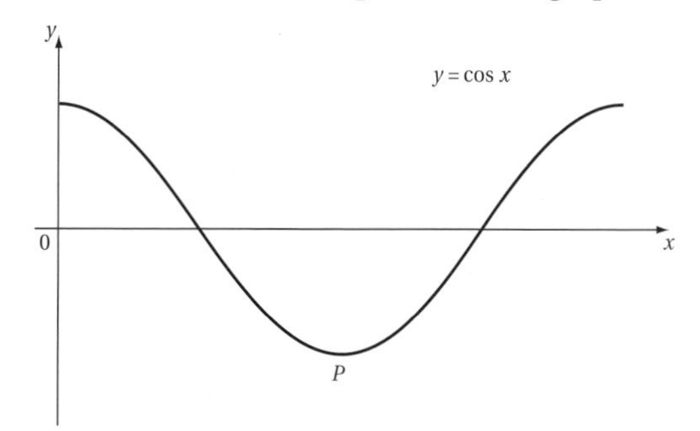

a Write down the coordinates of point P.

b Sketch, on separate diagrams, these graphs for $0° \leqslant x \leqslant 360°$.

i $y = \cos(x + 90°)$ **ii** $y = -\cos x$

iii $y = \cos \frac{x}{2}$

On each sketch label the image P' of point P with its coordinates.

11 a Starting with the graph of $y = \sin x$ sketch, on separate diagrams, these graphs for $0° \leqslant x \leqslant 360°$.

i $y = 2 \sin x - 1$ **ii** $y = -3 \sin x$

iii $y = 3 \sin(x - 90°)$

One of the graphs sketched in part **a** is identical to the graph of $y = k \cos x$.

b Identify this graph and state the value of k.

12 The diagram shows the graph of $y = \frac{1}{x}$ where $x \neq 0$.

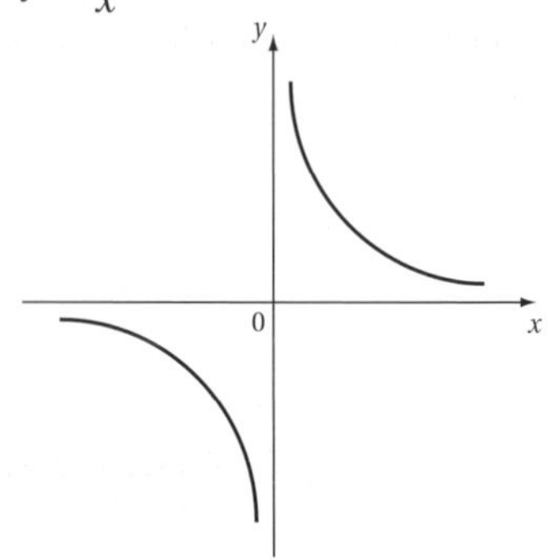

Sketch, on separate diagrams, these graphs. Show clearly the asymptotes and label them with their equations. Where appropriate, use the y-intercept to help to complete the sketch.

a $y = \frac{1}{x} + 3$ **b** $y = \frac{1}{x - 3}$

c $y = \frac{3}{x} + 1$ **d** $y = \frac{1}{x + 2} - 1$

e $y = \frac{2}{x - 4} + 3$ **f** $y = -\left(\frac{3}{x - 2}\right)$

13 The diagram shows the graph with equation $y = \frac{1}{x + a} - 1$ where a is a constant.

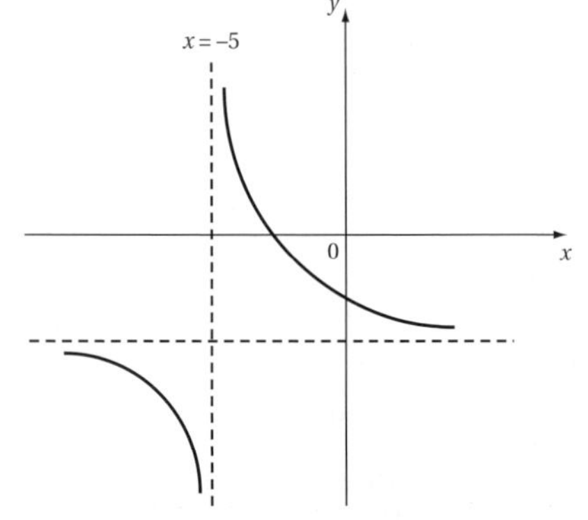

The dotted lines are asymptotes to the curve. The vertical asymptote has equation $x = -5$.

a State the value of a.

b Write down the equation of the horizontal asymptote.

c Find the coordinates of

i the y-intercept of this curve

ii the point where this curve intersects the x-axis.

14 By sketching a graph, or otherwise, find the equations of the asymptotes to the curves with these equations.

a $y = \frac{x + 1}{x}$

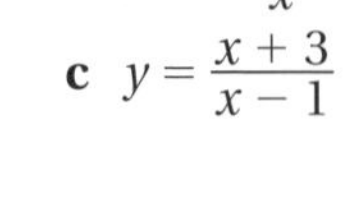

b $y = \frac{1 - 2x}{x}$

c $y = \frac{x + 3}{x - 1}$

Handy hint

Express each equation as the sum of two terms.

5.3 Intersection points of graphs

1 Find the coordinates of the points where these curves intersect the x-axis.

(a)

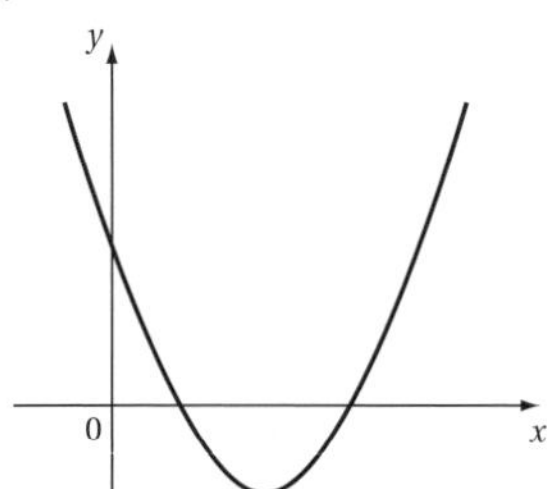

(b)

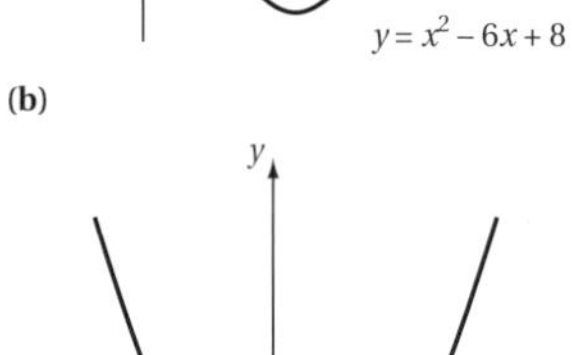
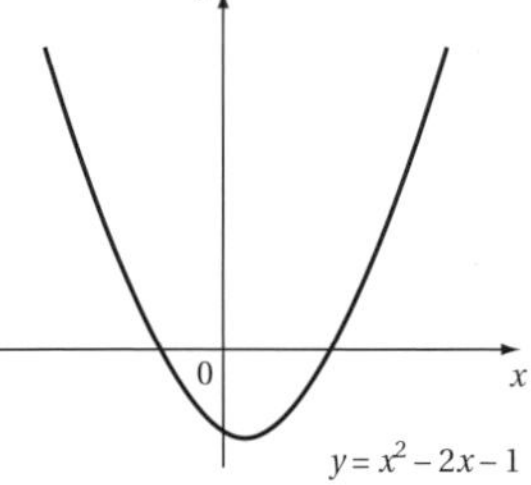

Give answers in simplified surd form where appropriate.

2 The diagram shows the curve $y = x^2 - 6x + 13$ and the line $y = 2x + 1$. The curve and line intersect at points A and B.

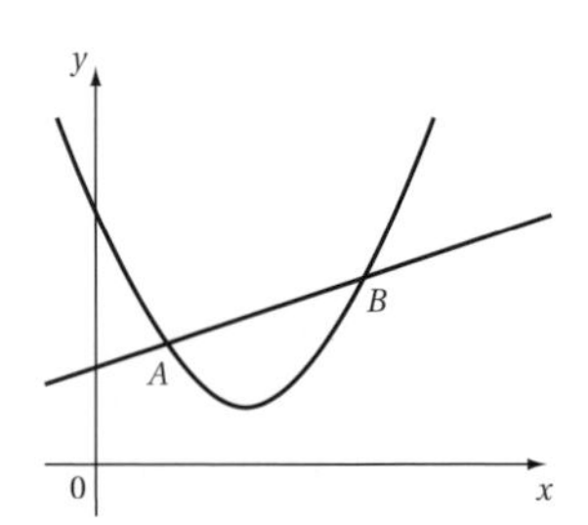

a Find the coordinates of A and B.

b Show that the distance $AB = 4\sqrt{5}$ units.

3 Find the coordinates of the points of intersection of these curves and lines. Give answers in simplified surd form where appropriate.

a Curve: $y = x^2 - 3x - 4$ Line: $y = 4 - x$

b Curve: $y = x^2 + 5x + 7$ Line: $y = 2x + 7$

c Curve: $y = x^2 + 6x + 3$ Line: $y = 4x + 4$

d Curve: $y = \frac{1}{6}x^2$ Line: $y - x + 1 = 0$

4 The diagram shows the curve $y = -x^2 + 7x - 6$ and the line $y = 4$. The curve crosses the x-axis at points A and B. The curve and line intersect at points C and D.

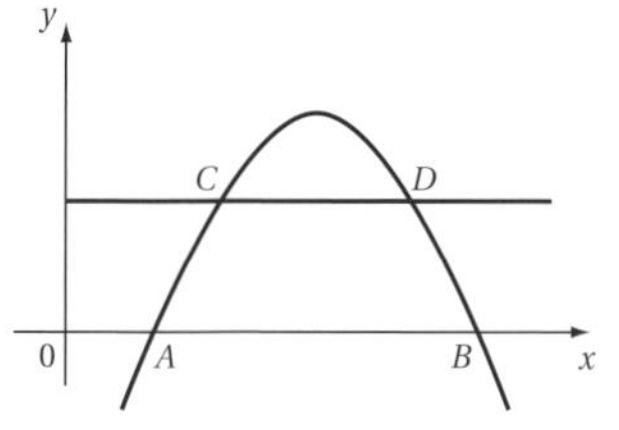

a Find the coordinates of A and B.

b Find the coordinates of points C and D.

c Hence find the area of the trapezium $ABCD$.

5 The diagram shows the curve $y = 2x^2 - 4x + 5$ and the line $y = 8x - 13$. The curve crosses the y-axis at point A. The line crosses the y-axis at point B.

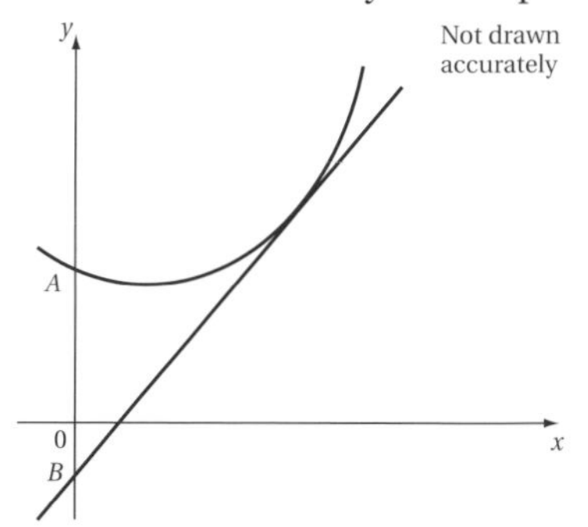

a Write down the coordinates of A and B.

b Show that this line and curve intersect at a single point, P, and give the coordinates of P.

c Show that triangle ABP has area 27 square units.

6 A curve has equation $y = x^2 - 4x + 7$ and a line has equation $y = 1 - 2x$.

a Use the method of substitution to show that the simultaneous equations $y = x^2 - 4x + 7$ and $y = 1 - 2x$ have no real solutions.

b What information about this curve and line does the result of part **a** give?

c Express $x^2 - 4x + 7$ in the form $(x - p)^2 + q$, where p and q are constants.

d Sketch, on the same diagram, the graph of $y = x^2 - 4x + 7$ and the line $y = 1 - 2x$.

7 **a** Express $x^2 + 6x + 13$ in completed square form.

b Sketch the graph of $y = x^2 + 6x + 13$. Label the vertex of your sketch with its coordinates.

c Express as an inequality the range of values of k such that the horizontal line $y = k$ does not intersect the curve $y = x^2 + 6x + 13$.

8 The line $y = k - 2x$, where k is a constant, intersects the curve $y = x^2 + 10x + 29$ at exactly one point P.

a Use this information to show that the equation $y = x^2 - 10x + 29$ is satisfied by exactly one value of x.

b By expressing $x^2 - 8x + 29 - k$ in completed square form, or otherwise, show that $k = 13$.

c Find the coordinates of point P.

9 The diagram shows the graph of $y = x^2 - 8x + 21$ and $y = -x^2 + 6x + 9$. The curves intersect at points A and B.

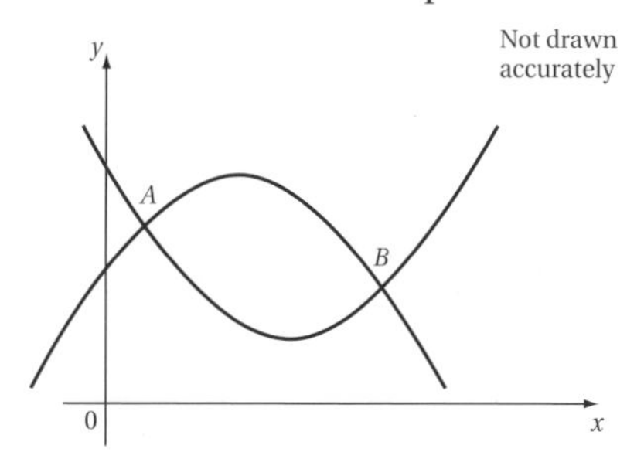

a Show that the x-coordinates of the points where these curves intersect satisfy the equation $x^2 - 7x + 6 = 0$.

b Find the coordinates of A and B.

The line L is the perpendicular bisector of AB.

c Show that the equation of L is $y = x + 8$.

d Find, to 1 decimal place, the coordinates of the points where L intersects the curve $y = -x^2 + 6x + 9$. You may use a calculator for this part of the question.

10 The diagram shows the graph of $y = x^2 - 6x + 13$ and the line $y = 2x + 6$. The line intersects the curve at points A and B.

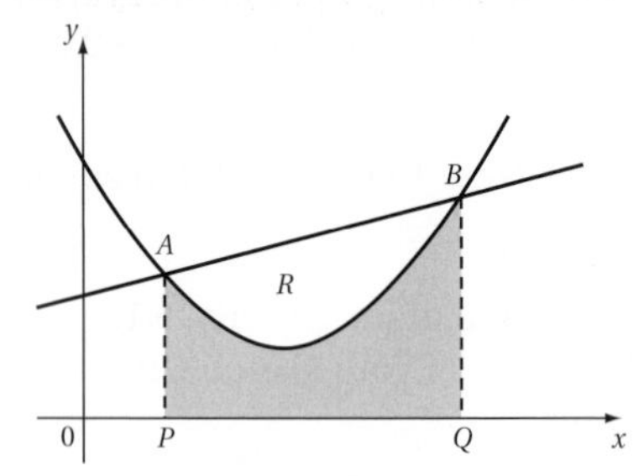

a Find the coordinates of A and B.

Points P and Q on the x-axis are such that the lines PA and QB are vertical. The shaded region bounded by the curve, the x-axis and the lines AP and BQ is 48 square units.

b Find the area of the region R between the curve and the line AB.

5.4 Three circle theorems

1 The diagram shows a circle with centre C and radius 5 cm. A tangent to the circle at point P is also shown. Q is a point on this tangent such that the distance $PQ = 12$ cm

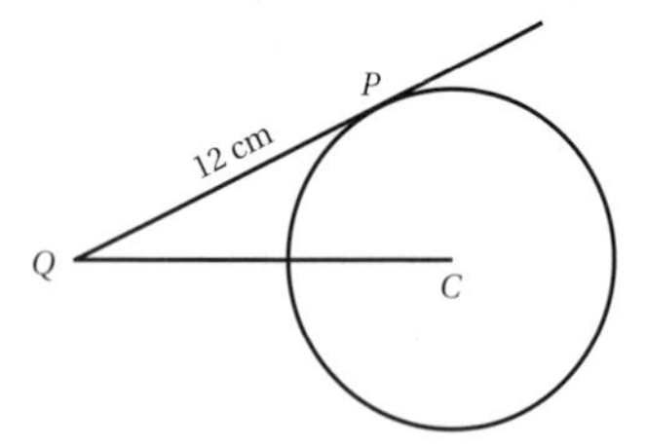

a Show that the distance $QC = 13$ cm.

b Determine whether the mid-point M of QC lies inside or outside this circle.

2 The diagram shows a circle with radius 10 cm. The line AB is a diameter of this circle. P is a point on this circle such that the distance $AP = 16$ cm.

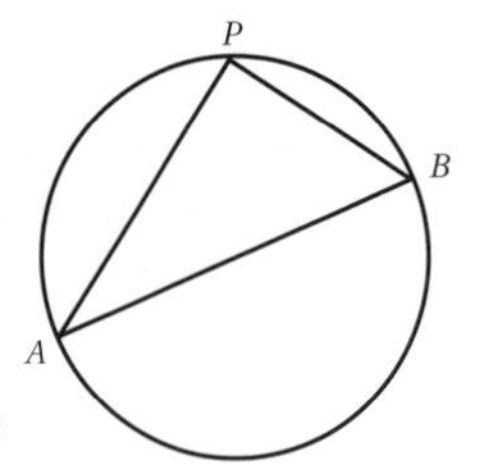

a Show that the distance $BP = 12$ cm.

b Find the area of triangle ABP.

3 The diagram shows a circle with centre C. The circle cuts the x-axis at points $A(2,0)$ and $B(8,0)$.

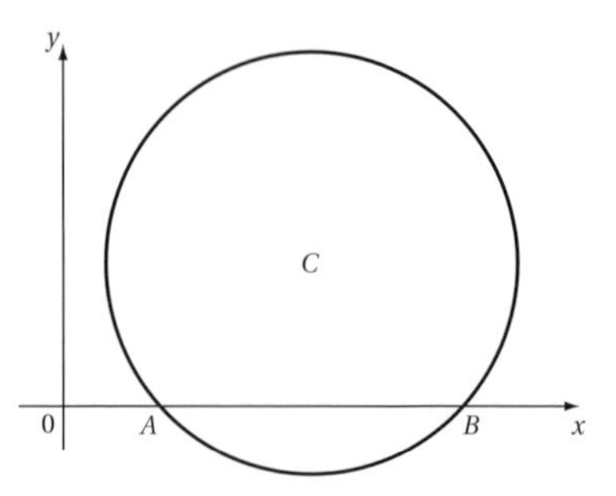

a Explain why the x-coordinate of C is 5.

It is given that the distance OC $= 4\sqrt{2}$ cm.

b Find the y-coordinate of C. Give your answer in surd form.

4 The diagram shows a circle with centre $C(5,3)$. Also shown is a tangent to the circle at point $P(2,9)$.

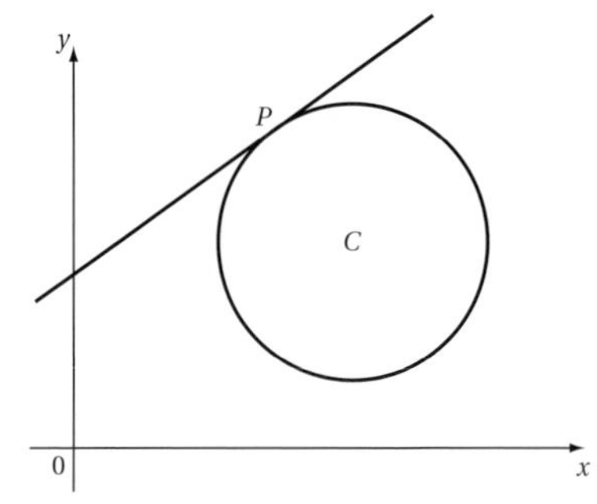

a Show that the gradient of the line CP is -2.

b Hence write down the gradient of this tangent.

c Find the equation of this tangent. Give your answer in the form $ay = bx + c$ where a, b and c are integers.

5 The diagram shows a circle with centre $C(8,4)$. The line L has equation $3y + x = 6$ and intersects this circle at points A and B, as shown.

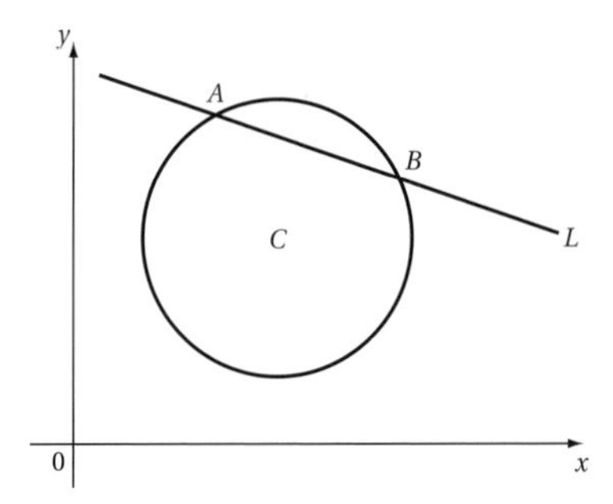

a Find the gradient of line L.

b Hence write down the gradient of the perpendicular bisector of AB.

c Find the equation of the perpendicular bisector of AB.

6 The diagram shows a circle with centre C(5,4). The tangent to this circle at the point P(8,6) has been drawn.

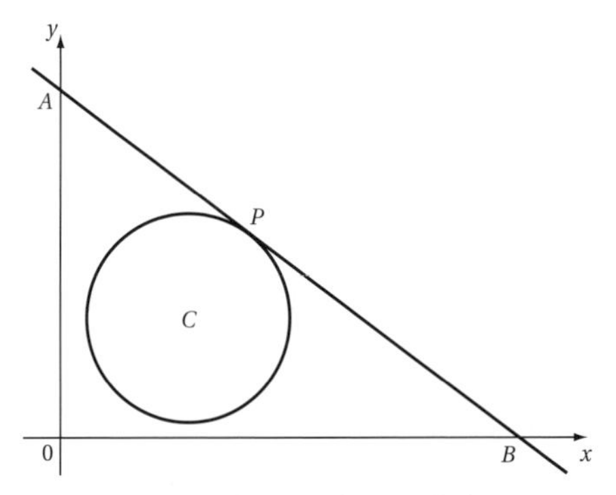

a Show that the gradient of this tangent is $-\frac{3}{2}$.

b Find the equation for this tangent. Give your answer in the form $ay + bx = c$ where a, b and c are integers.

c Find the coordinates of the points A and B where this tangent crosses the x and y axes.

d Show that the region inside triangle OAB excluding this circle has area $(108 - 13\pi)$ squared units.

7 The diagram shows a circle, centred at the origin, and with radius 5.

The line L_1 intersects this circle at the point $P(-5,0)$ and the point Q.

The line L_2 intersects this circle at the point Q and the point $R(5,0)$.

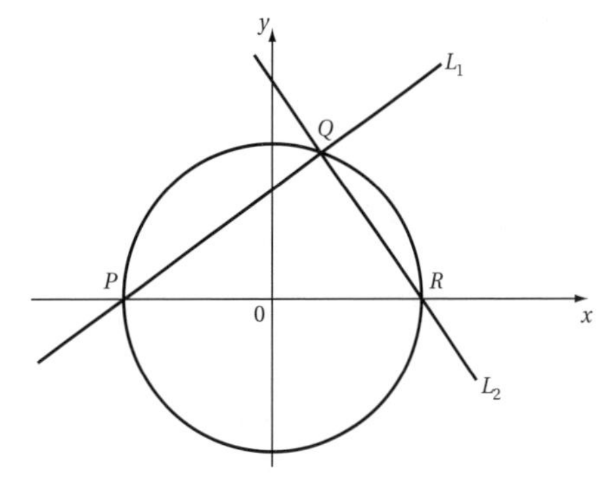

The equation of L_1 is $2y = x + 5$.

a Show that the equation of L_2 is $y = 10 - 2x$.

b Find the coordinates of point Q.

c By using the diameter PR, show that triangle PQR has area 20 square units.

d Hence, or otherwise, state the value of $QP \times QR$.

6 Practice: Trigonometry and triangles

6.1 Trigonometry and triangles

Unless told otherwise, use a calculator and give final answers to 3 significant figures.

1 Use right-angled trigonometry to find the value of x in these diagrams. All lengths are in centimetres.

a

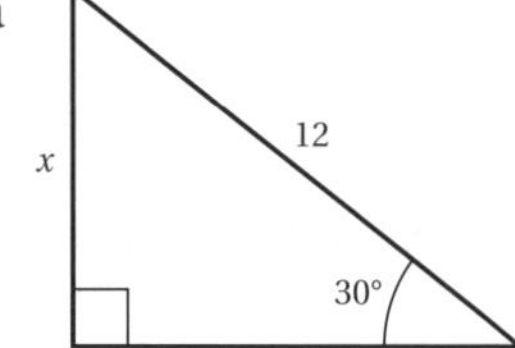

b

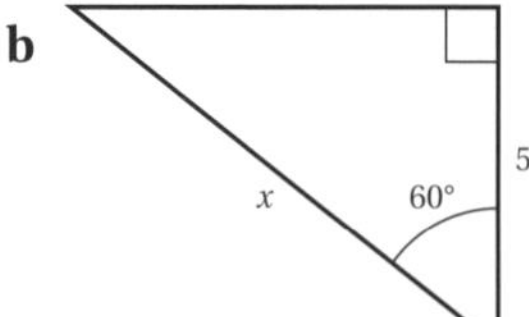

c

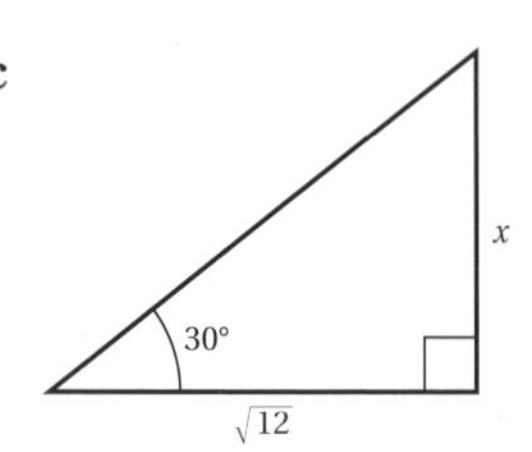

2 The diagram shows a right-angled triangle ABC. $AC = 2$ cm and $BC = 1$ cm. Angle $CAB = x$.

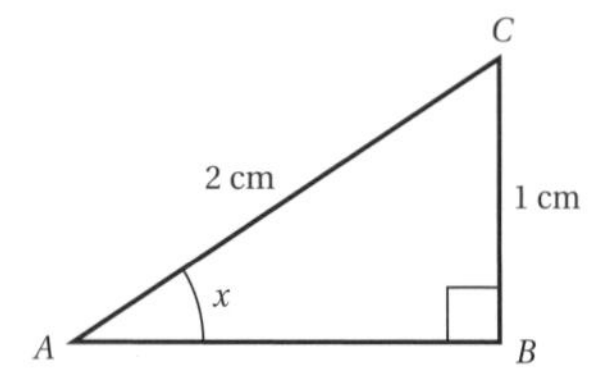

a Use right-angled trigonometry to show that $x = 30°$.

b Use Pythagoras' theorem to show that $AB = \sqrt{3}$.

c Use the triangle to find the **exact** values of

i cos 30° **ii** tan 60°.

Check each answer on a calculator.

3 **a** Use right-angled trigonometry to find the value of x in this diagram.

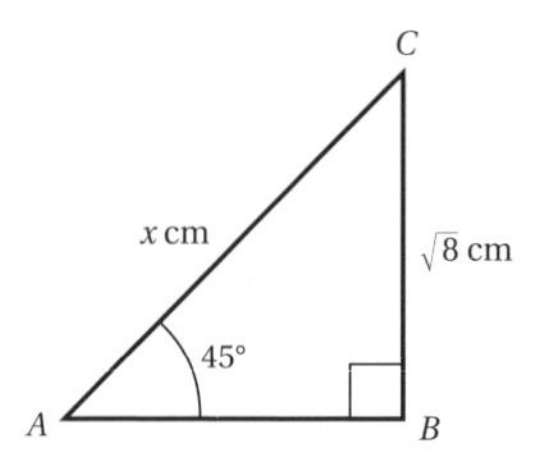

b Write down the length of the side AB.

c Use the triangle to find the value of

i tan 45°

ii cos 45°, giving your answer in simplified surd form.

Check each answer using a calculator.

4 Rearrange these expressions to make the required term the subject.

a $\frac{a}{\sin \hat{A}} = \frac{b}{\sin \hat{B}}$ for a

b $\frac{\sin \hat{A}}{a} = \frac{\sin \hat{C}}{c}$ for $\sin \hat{C}$

c $\frac{b}{\sin \hat{B}} = \frac{c}{\sin \hat{C}}$ for $\sin \hat{B}$

5 In these triangles, all lengths are in centimetres. Use the sine rule to find the length of the side indicated with a lowercase letter.

a

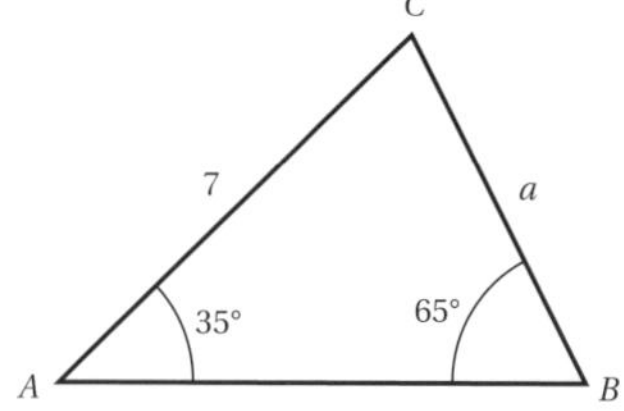

b

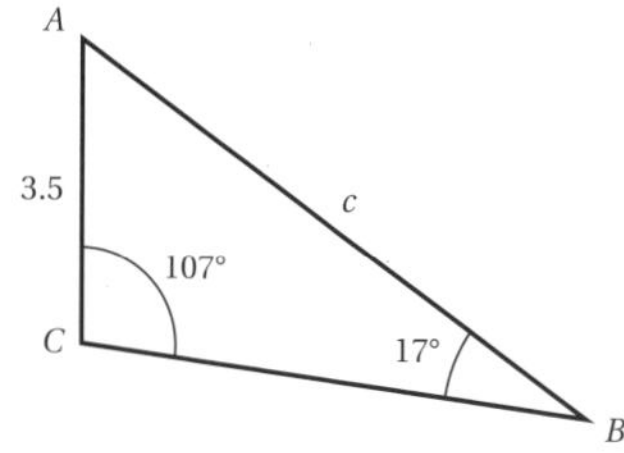

c

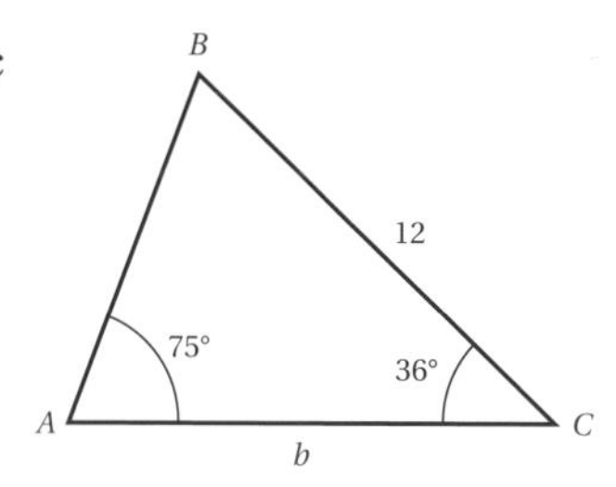

d

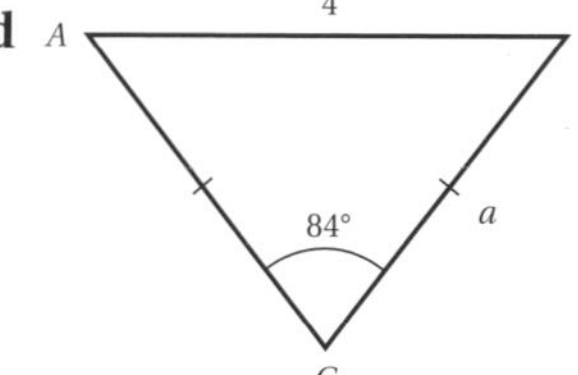

Handy hint

The dashes on sides AC and BC mean $AC = BC$.

6 The diagram shows triangle ABC. $AC = 6$ cm, $BC = 10$ cm and angle $BAC = 74°$.

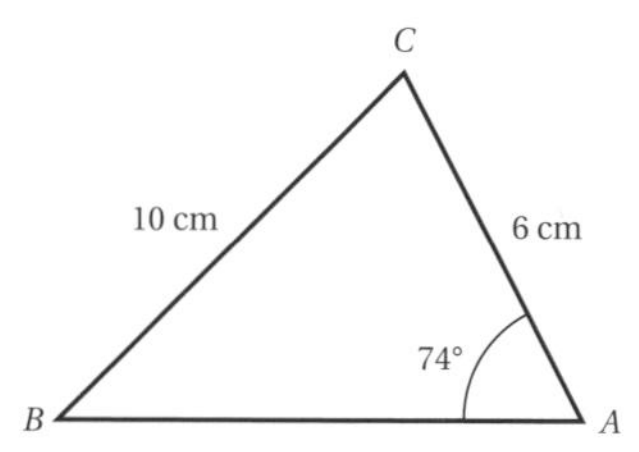

a Show that angle $CBA = 35.2°$ to 3 significant figures.

b Use the sine rule to find the length AB.

7 The diagram shows triangle ABC. $AB = x$, $AC = 2x$ and angle $ACB = 30°$

It is given that $\sin 30° = \frac{1}{2}$.

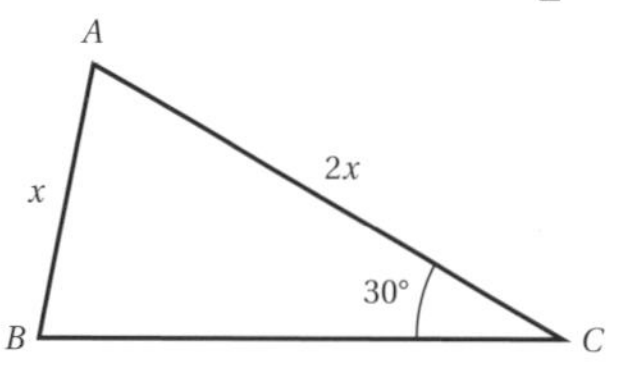

a Use the sine rule to show that $\sin \hat{B} = 1$.

b Hence show that triangle ABC is right-angled.

c Express the length of the side BC in terms of x, simplifying your answer as far as possible.

8 Rearrange these cosine rules to make the required term the subject.

a $c^2 = a^2 + b^2 - 2ab\cos\hat{C}$ for $\cos\hat{C}$

b $b^2 = a^2 + c^2 - 2ac\cos\hat{B}$ for $\hat{B}$

9 In these triangles, all lengths are in centimetres. Use the cosine rule to find the length of the side indicated with a lowercase letter.

a

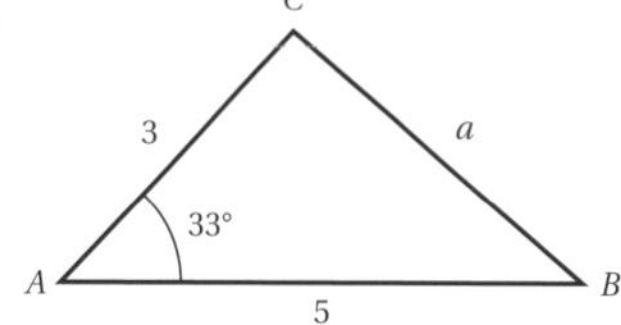

b

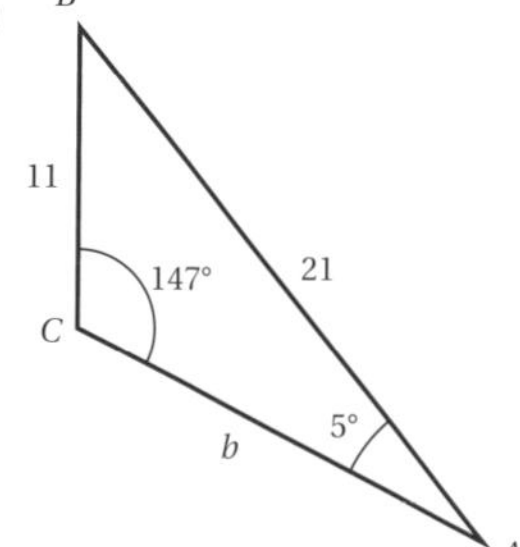

10 In any triangle, the largest angle is opposite the longest side.

a Use the cosine rule to find the largest angle in this triangle.

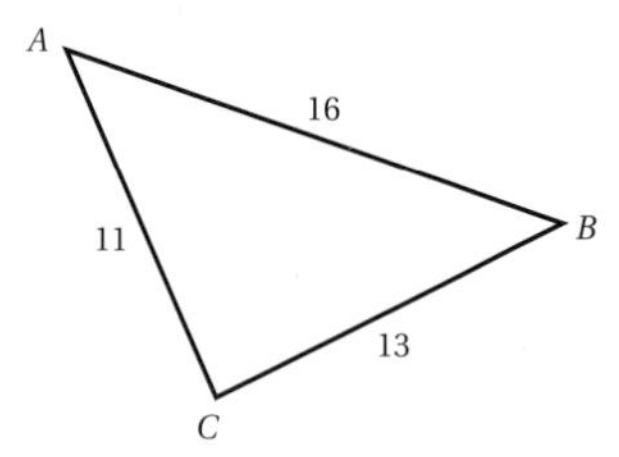

b Use any appropriate rules to find the other two angles of this triangle.

11 The diagram shows triangle *ABC*.
$AB = \sqrt{3}x$, $AC = 3x$ and angle BAC = 30°.

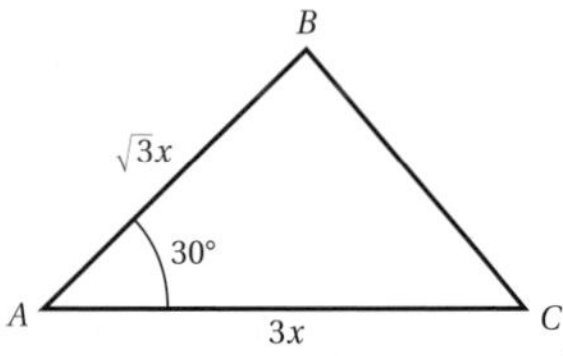

It is given that $\cos 30° = \frac{\sqrt{3}}{2}$.

a Show that $BC = \sqrt{3}x$.

Using properties of isosceles triangles, or otherwise,

b find angle *ABC*

c find the area of triangle *ABC*, giving your answer in the form $\frac{3}{4}\sqrt{k}x^2$ where k is an integer.

12 The diagram shows the triangle *ABC*, where *AB* = 7 cm, *AC* = 12 cm and angle *BAC* = 25°. Point *D* on *AC* is such that *AD* = 4 cm and angle *DBC* = 50°.

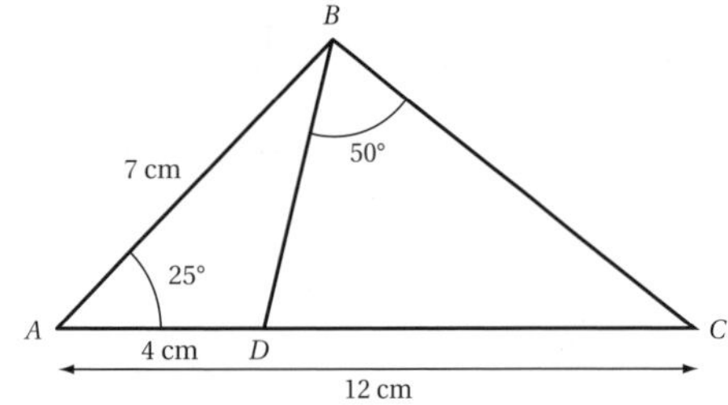

a Show that *BD* = 3.77 cm (to 3 significant figures).

b Find angle *C*.

c Hence, or otherwise, find angle *DBA*.

13 The diagram shows the sector of a circle with centre *C*. Points *A* and *B* lie on this circle. $DB = \sqrt{14}$ cm, where *D* is the mid-point of *AC*. $DC = x$ cm and angle *DCB* = 120°. It is given that $\cos 120° = -\frac{1}{2}$.

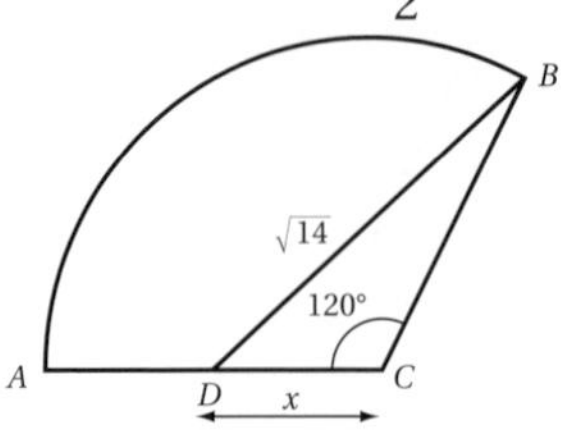

a Show that $x = \sqrt{2}$ cm.

b Find the perimeter of the curved shape *ADB*.

14 The diagram shows a quadrilateral *ABCD*. All lengths are in centimetres. *AB* = 10, *BC* = 18, *CD* = 12 and *DA* = 14.

Angle *ABD* = 27°.

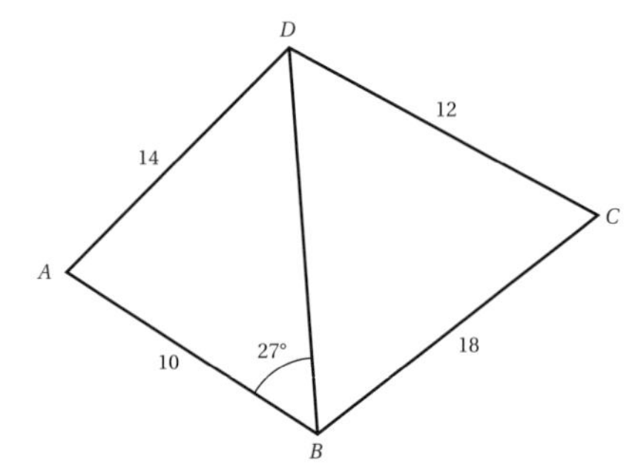

a Show that angle *ADB* = 18.9° (to 3 significant figures).

b Find angle *DCB*.

c Explain briefly why the points *A*, *B*, *C* and *D* cannot all lie on a common circle.

d Show that the diagonal *AC* has length 16 cm to the nearest centimetre.

6.2 The area of any triangle

Unless told otherwise, use a calculator and give final answers to 3 significant figures.

1 Find the area of each of these triangles.

a

b

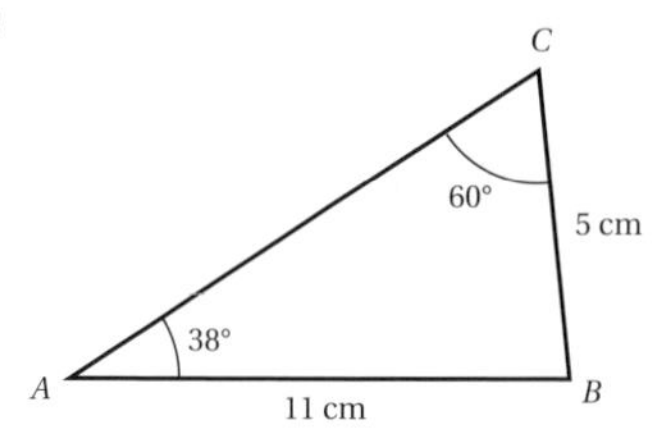

c

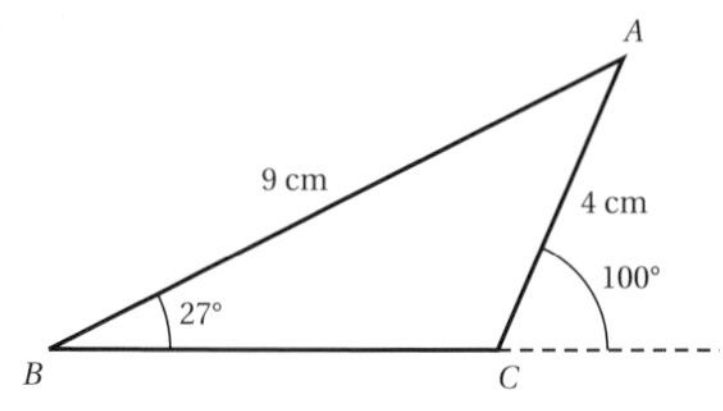

2 Find the area of this isosceles triangle.

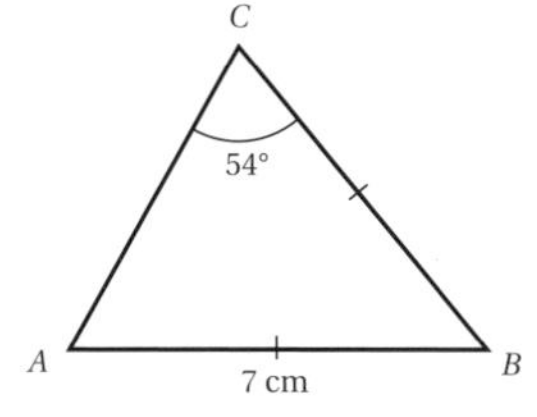

3 The diagram shows triangle PQR.
$PQ = 10$, $PR = 30$ and angle $PQR = 150°$.
All lengths are in centimetres.

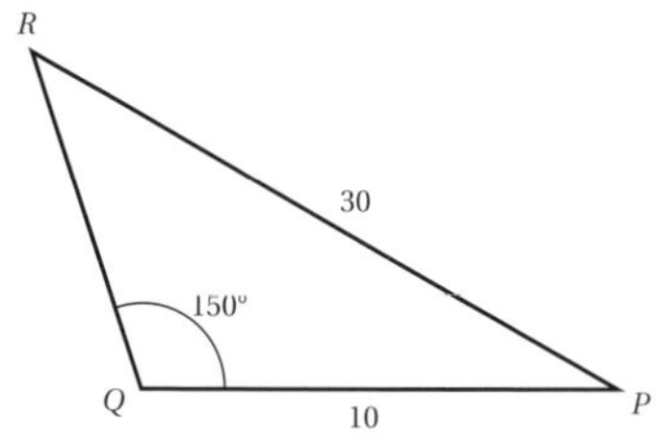

a Use the sine rule to show that $\sin \hat{R} = \frac{1}{6}$.

b Hence find angle QRP.

c Show that triangle PQR has area 52.3 cm^2 to 3 significant figures.

4 Find the area of an equilateral triangle whose perimeter is 15 cm.

5 In triangle ABC,
$AB = 7$ cm
$BC = 8$ cm
$AC = 9$ cm.

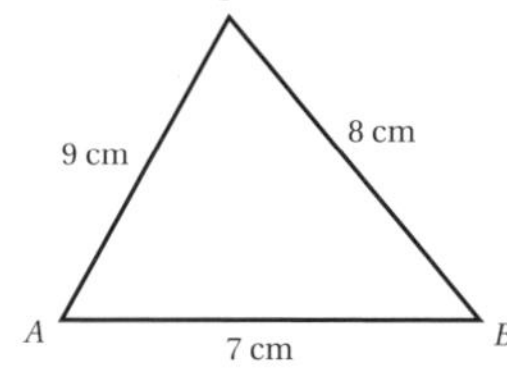

a Use the cosine rule to show that angle $ACB = 48.2°$ (to 3 significant figures).

b Hence find the area of this triangle.

6 The diagram shows a circle, radius 8 cm and centre at point C.
Points A and B on the circle are such that angle $ACB = 45°$.

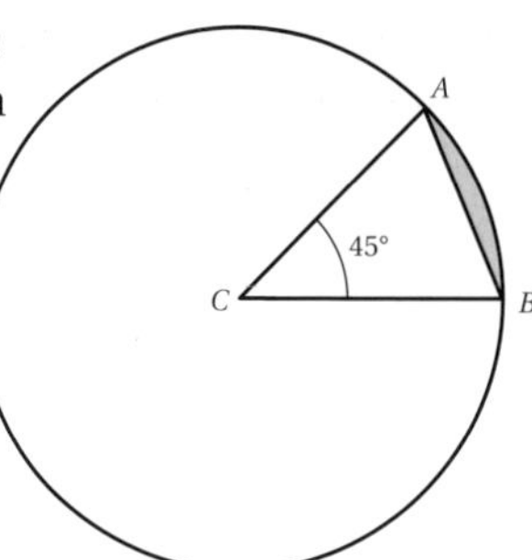

a Find the area of this circle. Leave π in your answer.

b Find the area of triangle ABC.

c Show that the shaded segment between this circle and the line AB has area 2.5 cm^2 (to 1 decimal place).

7 The triangles ABC and PQR shown in the diagram have equal areas.

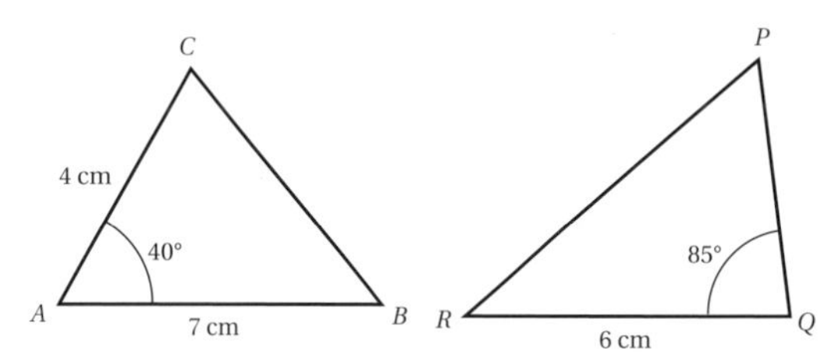

a Find the area of triangle ABC.

b Hence find the length of the side QP.

c Show that the perimeter of triangle PQR is 15.5 cm (to 3 significant figures).

8 The diagram shows the triangle PQR where $PQ = x$, $PR = 2x$ and angle $QPR = 30°$.
All lengths are in centimetres.

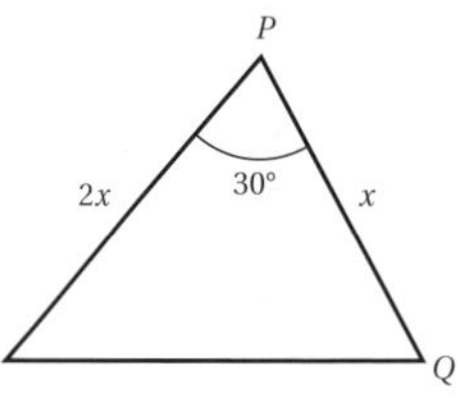

It is given that $\sin 30° = \frac{1}{2}$.

a Show that the area A of this triangle is given by the formula $A = \frac{1}{2}x^2$.

It is given that the area of this triangle is 18 cm^2.

b Find the length of the side PQ.

c Hence show that the base RQ of this triangle has length 7.44 cm (to 3 significant figures).

d Find the length of the shortest line from P to the side RQ.

9 The diagram shows the design of a badge in the shape of a sector ABC of a circle. The circle has centre A and radius 10 cm. Angle BAC is 85°.

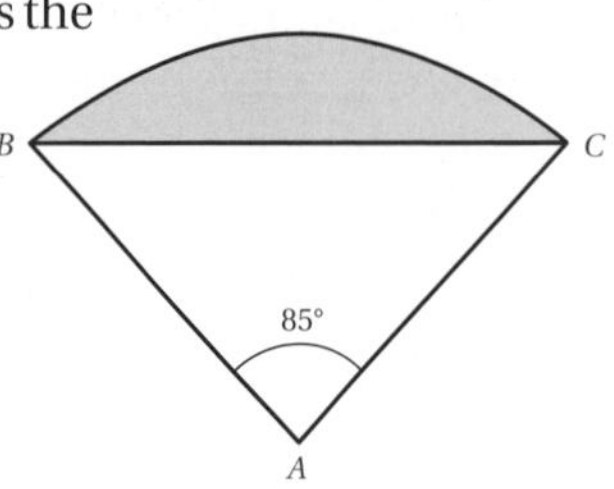

a Find, in square centimetres to 1 decimal place, the area of triangle ABC.

b Show that the area of the shaded region is 24.4 cm^2 to 1 decimal place.

10 A metal plate is formed by welding triangle ABC to a sector BCD of a circle. The circle has centre C and radius 7 cm. Angle BAC is 25° and angle BCD is 55°.

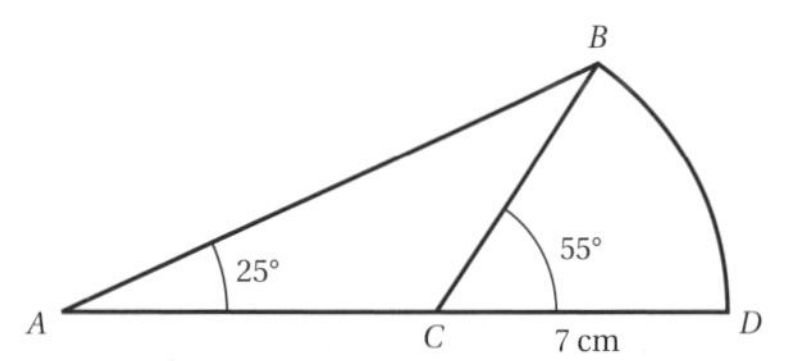

a Show that the length of AB is 13.6 cm, to 3 significant figures.

b Find the area of the plate.

c Find the perimeter of the plate.

11 The diagram shows a rectangle $ABCD$, where $AB = 4$ cm. Point E on BC is such that $BE = 3$ cm and point F on CD is such that $EF = \sqrt{2}$ cm and $AF = \sqrt{17}$ cm.

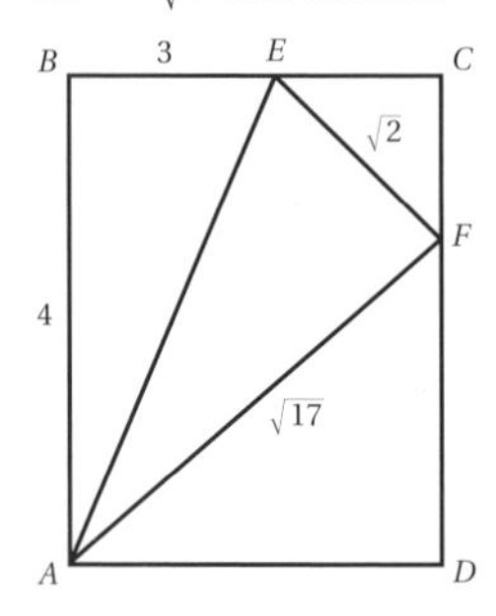

Not drawn accurately

a Show that angle $AEF = 45°$.

b Find the area of triangle AEF.

c Using calculator accuracy, show that $EC = \frac{1}{5}$.

d Hence find the area of triangle AFD.

6.3 Solving a trigonometric equation

Where appropriate, use a calculator and give answers to 1 decimal place.

1 The diagram shows the graph of $y = \sin x$ for $0° \leqslant x \leqslant 360°$.

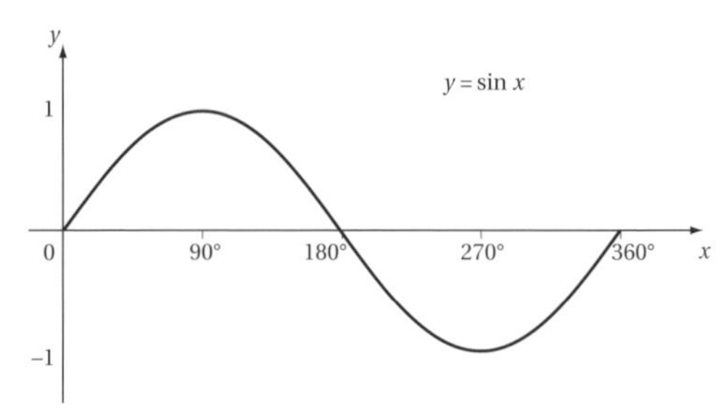

a Use this graph to write down the solutions to the equation $\sin x = 0$ for $0° \leqslant x \leqslant 360°$.

b Use your calculator to find one solution to the equation $\sin x = 0.6$.

c Use this graph to find another solution to the equation $\sin x = 0.6$ for $0° \leqslant x \leqslant 360°$.

2 The diagram shows the graph of $y = \cos x$ for $0° \leqslant x \leqslant 360°$.

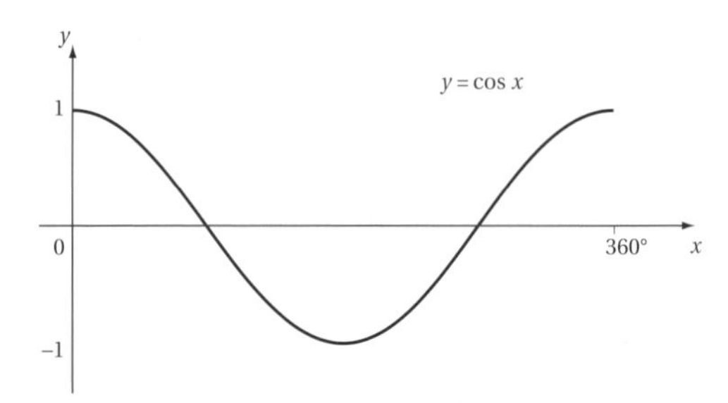

a Write down the solution of the equation $\cos x = -1$ for $0° \leqslant x \leqslant 360°$.

b Solve the equation $\cos x = -0.7$ for $0° \leqslant x \leqslant 360°$.

3 By using a calculator and the properties of the sine graph, solve these equations.

a $\sin x = 1$ for $0° \leqslant x \leqslant 360°$

b $\sin x = 0.25$ for $0° \leqslant x \leqslant 360°$

c $\sin x = 0.5$ for $0° \leqslant x \leqslant 540°$

4 By using a calculator and the properties of the cosine graph, solve these equations.

a $\cos x = 1$ for $0° \leqslant x \leqslant 360°$

b $\cos x = \frac{2}{3}$ for $0° \leqslant x \leqslant 360°$

c $\cos x = -\frac{1}{\sqrt{2}}$ for $0° \leqslant x \leqslant 720°$

5 The diagram shows the graph of $y = \sin x$ for $0° \leqslant x \leqslant 360°$.

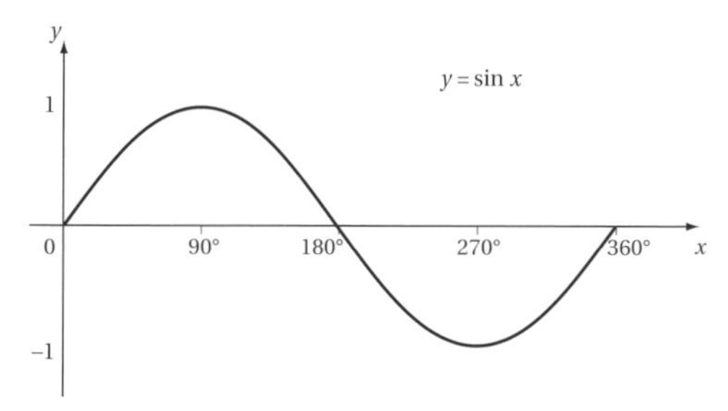

a Use your calculator to verify that $\sin 45° = \cos 45°$.

b On a copy of this diagram, sketch the graph of $y = \cos x$ for $0° \leqslant x \leqslant 360°$.

c Use the result of part **a** to write down a solution to the equation $\sin x = \cos x$. Illustrate this solution on your sketch.

d Use your sketch to find the other solution to the equation $\sin x = \cos x$ for $0° \leqslant x \leqslant 360°$. Verify your answer using a calculator.

6 The diagram shows triangle ABC where $AC = 8$ cm, $BC = 7$ cm and angle $CAB = 30°$. The triangle has been drawn so that angle ABC is acute.

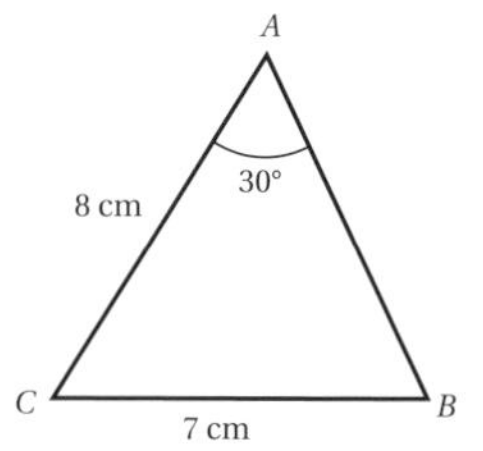

a Use the sine rule to show that $\sin \hat{B} = \frac{4}{7}$.

b Hence find the acute angle B.

c Find an obtuse angle which satisfies the equation $\sin x = \frac{4}{7}$.

d Using your answer to part **c**, sketch a triangle PQR, not congruent to triangle ABC, for which $PR = 8$ cm, $QR = 7$ cm and angle $RPQ = 30°$.

7 Solve these equations for $0° \leqslant x \leqslant 360°$.

a $2 \sin x = 1$ **b** $3(1 + \cos x) = 4$

c $3 + \sqrt{2}\sin x = 4$ **d** $\frac{3 - 4\cos x}{2} = 1$

8 a Show, with the aid of a sketch, that $\sin(x + 90°) = \cos x$ for all values of x.

b Hence solve the equation $\sin(x + 90°) = 0.2$ for $0° \leqslant x \leqslant 360°$.

c Solve, using a similar method, the equation $\cos(x + 90°) = -0.75$ for $0° \leqslant x \leqslant 360°$.

9 a Use your calculator to find a solution of the equation $\sin x = -0.5$.

Handy hint

For part **a**, illustrate this solution on a sketch of $y = \sin x$.

b Use the graph of $y = \sin x$ to find the smallest positive solution to the equation $\sin x = -0.5$.

c Find the two smallest positive solutions to the equation $\sin x = -\frac{4}{5}$.

10 a Show, with the aid of a sketch, why the equation $\sin x = \frac{5}{4}$ has no solutions.

b Solve, where possible, these equations for $0° \leqslant x \leqslant 360°$.

i $2 \cos x = 3$ **ii** $4 \sin x + 3 = -1$

iii $\frac{\sin x}{3} = -0.334$ **iv** $\frac{3}{\cos x} - 1 = 3$

11 a Describe the transformation which maps the graph of $y = \sin x$ to the graph of $y = \sin ax$, where a is a positive constant.

b Solve the equation $\sin x = 0.4$ for $0° \leqslant x \leqslant 360°$.

c Hence, or otherwise, solve these equations.

i $\sin 2x = 0.4$ for $0° \leqslant x \leqslant 180°$

ii $\sin 4x = 0.4$ for $0° \leqslant x \leqslant 90°$

iii $\sin \frac{3x}{2} = 0.4$ for $0° \leqslant x \leqslant 240°$

12 a Solve the equation $\cos x = -\frac{3}{8}$ for $0° \leqslant x \leqslant 360°$.

b Hence, or otherwise, solve these equations.

i $\cos 3x = -\frac{3}{8}$ for $0° \leqslant x \leqslant 120°$

ii $8 \cos 2x + 3 = 0$ for $0° \leqslant x \leqslant 180°$

13 Solve these equations for $-360° \leqslant x \leqslant 360°$.

a $\cos x = 0.28$ **b** $\sin x = 0.65$

c $\sin x = -0.7$ **d** $5 \cos x + 3 = 1$

7 Practice: Sequences and summations

7.1 Arithmetic sequences

1 **a** Find an expression for the nth term, u_n, of these arithmetic sequences. Simplify each answer as far as possible.

i $5, 8, 11, \ldots$ **ii** $-5, -3, -1, \ldots$

iii $12, 8, 4, \ldots$ **iv** $1, \frac{3}{2}, 2, \ldots$

v $\frac{11}{2}, 4, \frac{5}{2}, \ldots$ **vi** $\frac{11}{12}, \frac{13}{6}, \frac{41}{12}, \ldots$

b For each answer to part **a** find the value of u_{10}.

2 Which of these sequences are definitely **not** arithmetic? Justify each answer.

a $3, 0, -3, 0, \ldots$ **b** $4, 13, 22, 31 \ldots$

c $1, \frac{1}{2}, \frac{1}{3}, \frac{1}{4}, \ldots$ **d** $7, 7, 7, 7, \ldots$

e $0, \sqrt{2}, \sqrt{8}, \sqrt{18}, \ldots$ **f** the odd prime numbers in ascending order.

3 An arithmetic sequence begins 7, 11, 15 …

a Find an expression for u_n, the nth term of this sequence.

b Find the value of u_{20}.

c Which term in this sequence has value 171?

d Explain why 101 will not be a term in this sequence.

4 An arithmetic sequence begins $9, p, 15, \ldots$ where p is a constant. The nth term of this sequence is u_n.

a Find the value of p.

b Find an expression for u_n. Simplify your answer as far as possible.

c Explain why every term in this sequence is a multiple of 3.

d Which term in this sequence has value 201?

5 An arithmetic sequence begins 10, 14, 18, 22, …

a Find an expression for u_n, the nth term of this sequence.

b Find the value of the first term which is greater than 131.

c Prove that the sum of any pair of consecutive terms of this sequence is a multiple of 8.

6 An arithmetic sequence begins $a, 11, b, 25, \ldots$ where a and b are constants. The nth term of this sequence is u_n.

a Find the value of a and the value of b.

b Find an expression for u_n. Simplify your answer as far as possible.

c Find the terms of this sequence which are square numbers less than 100.

7 The nth term of an arithmetic sequence is u_n. The 1st term is 93 and the 10th term is 48. The common difference is d.

a Find the value of d.

b Which is the first term less than 60?

c Find the smallest positive term in this sequence.

8 In the 'Quid Pro Quo' savings scheme, an initial investment of £2000 grows by £50 per month, so that after the first month the value of the investment is £2050, after the 2nd month its value is £2100, and so on. The value (in £) of the investment after n months is u_n.

a Find an expression for u_n.

b Find the value of the investment after one year.

c At the end of which month will the investment have doubled from its initial value?

9 Jane buys a new car for £9100. Each month the car depreciates (i.e. loses value) at a rate of £100 per month, so that after one month the car is worth £9000, after two months it is worth £8900 and so on. The value (in £) of the car after n months is u_n.

a Show that $u_n = 100(91 - n)$.

When the value of the car first falls below £1000, Jane sells it for scrap.

b For how many whole months did Jane own the car?

Jane then buys another car. The value (in £) u_n of this car after depreciating for n months is given by the expression $u_n = 5(2200 - 23n)$.

c Find

i the value of this car when it was new

ii the rate at which this car depreciates.

10 Tom is revising for some exams to be taken in four weeks' time. He draws up a revision schedule. The arithmetic sequence 1.5, 2, 2.5, … describes the number of hours he plans to revise on the 1st, 2nd, 3rd etc day of the schedule.

a Find an expression for the number of hours Tom plans to revise on the nth day of his schedule.

b On which day in the schedule does Tom plan to revise for 5 hours?

c How many hours does Tom plan to revise for on the 28th day of the schedule? Comment on whether you think this answer is realistic.

d Find the **total** number of hours Tom plans to spend revising from the 1st to the 7th day of the schedule (inclusive).

11 An arithmetic sequence begins $a, a + d, a + 2d, \ldots$ where a is the first term and d is the common difference.

The 3rd term of this sequence is 8.

a Use this information to write down an equation involving a and d.

The 14th term of this sequence is 63.

b Use this information to write down another equation involving a and d.

c By solving these equations simultaneously, or otherwise, show that $d = 5$ and find the value of a.

d Hence find the value of the 50th term of this sequence.

12 An arithmetic sequence has first term a and common difference d. The nth term of this sequence is u_n.

a Write down an expression for u_n in terms of a and d.

The 5th term of this sequence is three times the 2nd term.

b Use this information to show that $d = 2a$.

The third term of this sequence is 10.

c Find the value of a and the value of d.

d Explain why no term in this sequence is exactly divisible by 4.

13 **a** Show that $4k^2$, $(2k + 1)(2k - 1)$, $2(2k^2 - 1)$, where k is any constant, form the first three terms in an arithmetic sequence.

b Find, in terms of k, the 5th term of this sequence. Factorise your answer as far as possible.

Given that the 10th term of this sequence is 0,

c find the possible values of k.

7.2 Recurrence relations and sigma notation

1 In these questions, the nth term of each sequence is u_n. For each sequence, find the value of u_2, u_3 and u_4.

a $u_1 = 2$, $u_{n+1} = u_n + 3$

b $u_1 = 3$, $u_{n+1} = 2u_n$

c $u_1 = -1$, $u_{n+1} = 3u_n + 4$

d $u_1 = 5$, $u_{n+1} = 4 - u_n$

e $u_1 = 0$, $u_{n+1} = 3 - 2u_n$

f $u_1 = 2$, $u_{n+1} = 4u_n - 6$

2 A sequence $u_1, u_2, u_3, \ldots$ is defined by

$u_1 = 3$

$u_{n+1} = -u_n$.

a Find the value of u_2, u_3 and u_4.

b Hence state the value of u_{2011}.

c State the value of $u_{n+1}u_n$ for any positive integer n.

3 A sequence $u_1, u_2, u_3, \ldots$ is defined by

$u_1 = 6$

$u_{n+1} = 3 + u_n$.

a Find the values of u_2 and u_3.

b State the name given to this type of sequence.

Handy hint
Refer to 7.1.

c Find an expression for u_n in terms of n only.

d Hence, or otherwise, find the value of u_{24}.

4 A sequence $u_1, u_2, u_3, \ldots$ is defined by

$u_1 = k$

$u_{n+1} = 2u_n + 7$, where k is a constant.

a Find an expression for u_2 in terms of k.

It is given that $u_2 = 3$.

b Show that $k = -2$.

c Find the value of u_4.

5 A sequence $u_1, u_2, u_3, \ldots$ is defined by

$u_1 = 4$

$u_{n+1} = 2u_n + k$, where k is a constant.

a Find an expression for u_2 in terms of k.

b Hence show that $u_3 = 3k + 16$.

Given that $u_3 = 7$,

c find the value of k.

d Find the value of u_5.

6 A sequence $u_1, u_2, u_3, \ldots$ is defined by

$u_1 = 2$

$u_{n+1} = ku_n + 4$, where k is a constant.

a Find an expression for u_2 in terms of k.

b Hence show that $u_3 = 2(k^2 + 2k + 2)$.

Given that $u_3 = 34$,

c find the two possible values of k.

d Hence find the two possible values of u_4.

7 Each of these sequences can be written in the form

$u_1 = 1$

$u_{n+1} = au_n + b$ for constants a and b.

For each sequence,

i find the value of a and the value of b,

ii find the value of u_4.

a $1, 4, 13, \ldots$ **b** $1, 5, -3, \ldots$

c $1, 2, \frac{5}{2}, \ldots$

8 Evaluate these sums.

a $\sum_{n=1}^{4} u_n$ where $u_n = 3n - 1$

b $\sum_{n=1}^{5} u_n$ where $u_n = 6 - 2n$

c $\sum_{n=1}^{4} u_n$ where $u_n = 2^n$

d $\sum_{n=1}^{5} u_n$ where $u_n = \frac{60}{n}$

9 a Find the nth term of the arithmetic sequence 2, 6, 10, ...

The Nth term of this sequence is 70.

b Find the value of N.

c Hence express in sigma notation the sum of the first N terms of this sequence. You should **not** attempt to evaluate this sum.

10 Write out in full, and hence evaluate, each of these sums.

a $\sum_{n=1}^{4} n^2$ **b** $\sum_{n=1}^{4} n(2n - 1)$

c $\sum_{n=1}^{3} \frac{1}{2}(3^{n-1})$ **d** $\sum_{n=1}^{4} \frac{n + 12}{n}$

e $\sum_{n=1}^{5} (-2)^n$

f $\sum_{n=4}^{7} \sqrt{n^2 - 2n + 1}$

Handy hint
In **f** use the **positive** square.

11 The nth term of the arithmetic sequence 30, 27, 24, ... is u_n.

a Find an expression for u_n.

b Which term of this sequence has value 0?

Given that $\sum_{n=1}^{N} u_n = 0$,

c find the value of N.

12 A sequence $u_1, u_2, u_3, \ldots$ is defined by

$u_1 = 1$

$u_{n+1} = 2(3u_n - 1)$.

a Show that $u_4 = 130$.

b Find the value of $\sum_{n=1}^{4} u_n$.

13 A sequence $u_1, u_2, u_3, \ldots$ is defined by

$u_1 = 2$

$u_{n+1} = au_n - 1$ where a is a constant.

It is given that $u_2 = 5$.

a Show that $a = 3$.

b Find the value of $\sum_{n=2}^{5} u_n$.

8 Practice: Introducing differentiation

8.1 Estimating the gradient of a curve

You may use a calculator in these questions.

1 The diagram shows part of the curve $y = x^2$ which passes through the points $A(1,1)$, $P(2,4)$, and $B(3,9)$. The tangent to the curve at P has been drawn.

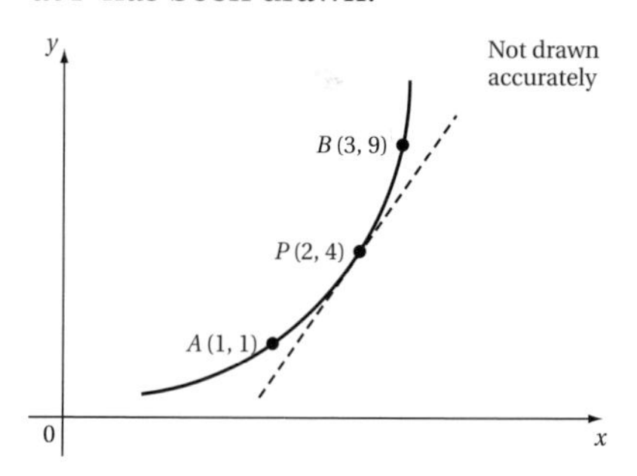

a Find the gradient of PB.

b Find the gradient of AP.

The gradient of the tangent to the curve at P is m, where m is an integer.

c Use your answers to **a** and **b** to write down the value of m.

2 The diagram shows part of the curve $y = 4x^2 + 3$ which passes through the point $P(1,7)$.

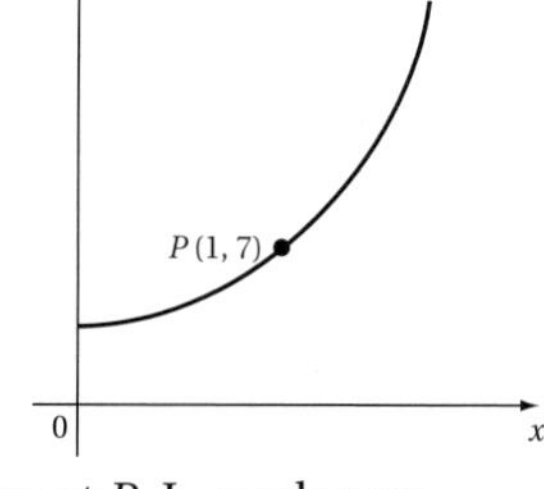

a Use each of these points to find an estimate for the gradient of the curve at P. In each case, state whether the answer gives an under- or over-estimate.

i $Q(0.5,4)$ **ii** $R(1.5,12)$

iii the point S where $x = \sqrt{2}$, giving your answer to 3 significant figures.

b State, with a reason, which of the answers found in part **a** gives the best estimate for the gradient of the curve at P.

3 The diagram shows the curve $y = x^2 - 4x + 7$ which passes through the points $P(3,4)$ and $Q(4,7)$. Point R is the vertex of this curve.

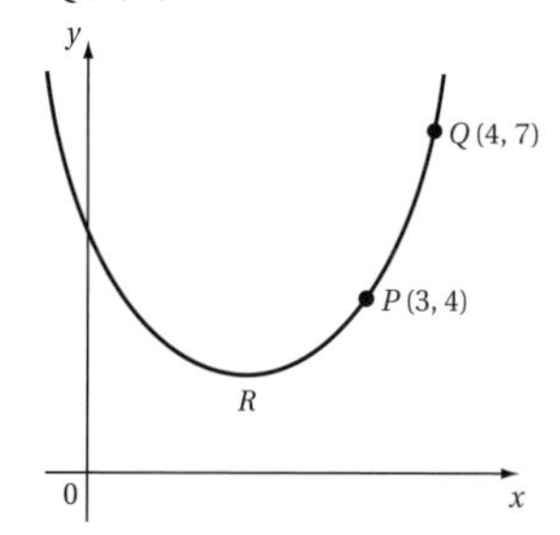

a Express $x^2 - 4x + 7$ in the form $(x - a)^2 + b$, where a and b are constants.

b Hence write down the coordinates of point R.

c Write down the gradient of the curve at point R.

The gradient of the curve at point P is m, where m is an integer.

d By finding the gradient of PQ and the gradient of RP, show that $m = 2$.

e Write down the gradient of this curve at the point where $x = 1$.

Handy hint
Use the symmetry of the curve.

4 By calculating $Grad_{PQ}$, where Q is a point on the given curve, obtain an estimate for the gradient of each curve at point P.

a curve: $y = 2x^2 - x + 4$, point $P(2,10)$, point $Q(2.5,14)$

b curve: $y = 3x - 4x^2$, point $P(2,-10)$, point $Q(1.75,-7)$

c curve: $y = 2x^2 + 1$, point $P(-2,9)$, point Q with x-coordinate -1.5

d curve: $y = \sqrt{2x^2 + 7}$, point P has x-coordinate 1, point Q has x-coordinate 3

5 The diagram shows part of the curve $y = \frac{1}{4}x^2 + 5$ which passes through the point P with x-coordinate 3.

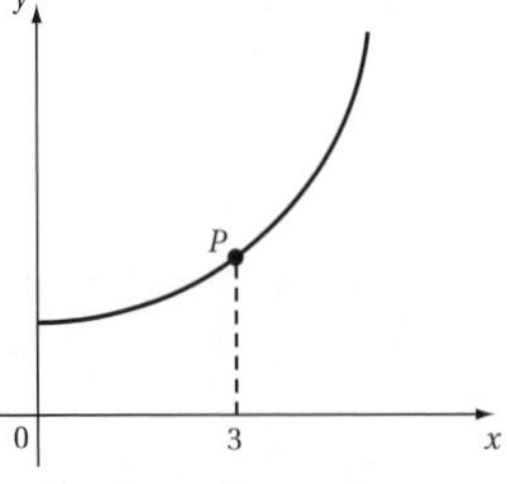

By choosing a suitable pair of points on this curve either side of P, show that the gradient of the curve at P is 1.5, correct to 1 decimal place.

6 The diagram shows part of the curve $y = 2 - \frac{3}{x}$ which passes through the point P with x-coordinate 2.

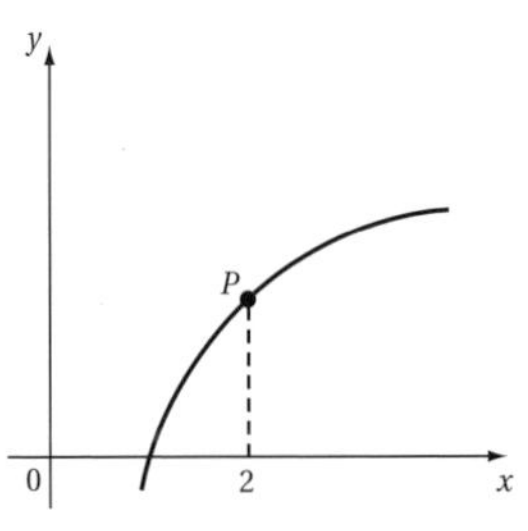

By choosing a suitable pair of points on this curve either side of P, determine the gradient of the curve at point P correct to 2 decimal places.

7 The diagram shows part of the curve $y = 2x^2$. The line $y = 5x - 3$ intersects this curve at points P and Q. The dotted line is the tangent to the curve at P. Point 0 is the origin.

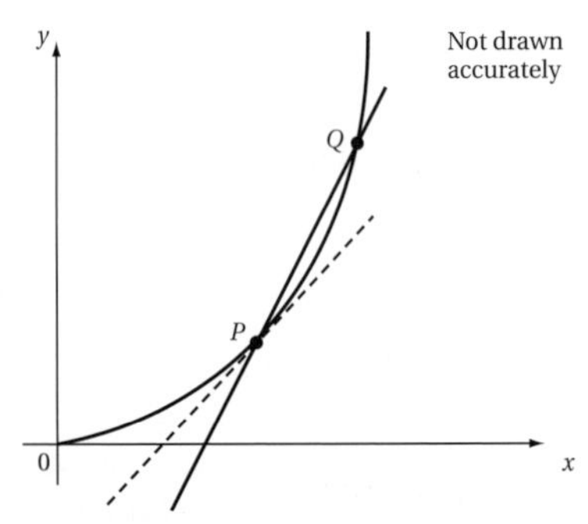

The equation of the tangent to the curve at P is $y = mx + c$, where m is an integer and $c < 0$.

a By comparing the tangent at P with the line PQ, explain why $m < 5$.

b Use algebra to find the coordinates of P and Q.

The tangent at P intersects the line OQ at a point which has a positive x-coordinate.

c Explain why $m > 3$.

d Write down the value of m and hence find the equation of the tangent to the curve at P.

8.2 Differentiation

1 The diagram shows the curve $y = 3x^2 + 1$. The point $P(x, 3x^2 + 1)$ is a general point on this curve. The point $Q(q, 3q^2 + 1)$ is a point on this curve which is close to P.

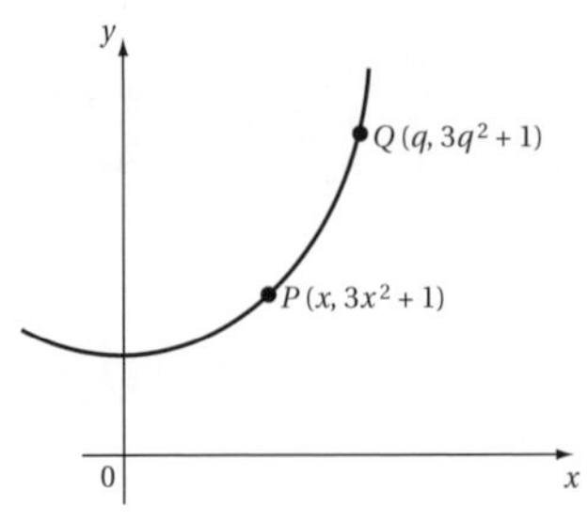

a Show that $Grad_{PQ} = 3(q + x)$.

b Hence find an expression for $\frac{dy}{dx}$ in terms of x.

2 The diagram shows the curve $y = x^2 - 3x + 4$ which passes through the points P and Q. Point P has coordinates $(x, x^2 - 3x + 4)$. The x-coordinate of point Q is q.

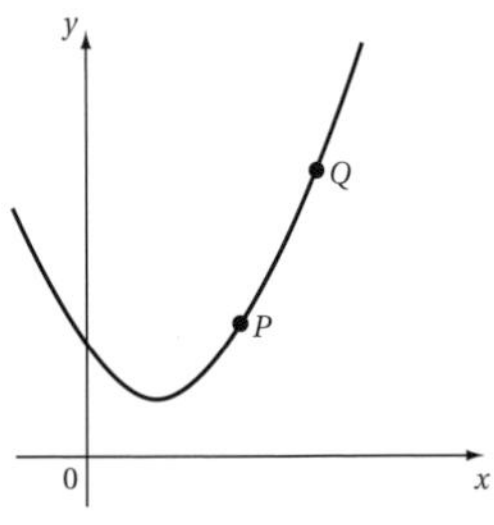

a Write down the y-coordinate of the point Q in terms of q.

b Show that $Grad_{PQ} = q + x - 3$.

c Hence find an expression for $\frac{dy}{dx}$ in terms of x.

3 **a** Factorise $q^2 - x^2$.

b Hence fully factorise $q^4 - x^4$.

c Hence show from first principles that the curve with equation $y = x^4$ has gradient equation $\frac{dy}{dx} = 4x^3$.

Questions 4 to 13 should be answered using the rules of differentiation.

Do **not** use first principles for these questions.

4 Find an expression for $\frac{dy}{dx}$ for the curves with these equations.

a $y = x^2 + 3x + 1$ **b** $y = 5x^2$

c $y = -3x^2 + 2$ **d** $y = \frac{1}{2}x^2 + 5x$

e $y = 4x - 2x^2$ **f** $y = 3 - 7x - 3x^2$

5 By first expanding the brackets in each of these equations, find $\frac{dy}{dx}$.

a $y = (x + 2)(x + 3)$ **b** $y = (2x + 1)(x - 3)$

c $y = x(4 - 3x)$ **d** $y = (3x - 4)(2 - 5x)$

e $y = (2x + 3)^2$ **f** $y = \left(3 - \frac{1}{2}x\right)^2$

6 Find the gradient of these cubic curves at the point indicated.

a $y = x^3 + 2x^2 + 1$ at the point where $x = 1$

b $y = 2x^3 - 3x - 4$ at the point where $x = \frac{1}{2}$

c $y = \frac{5}{3}x^3 + \frac{3}{2}x^2 + 2x$ at the point where $x = -2$

d $y = 7 - 3x - \frac{x^3}{6}$ at the point $\left(2, -\frac{1}{3}\right)$

7 By expressing each equation in an appropriate form, find $\frac{dy}{dx}$.

Factorise each answer as fully as possible.

a $y = x^3(x + 1)$ **b** $y = x(x^2 - 12)$

c $y = x(x^2 - 6x + 12)$ **d** $y = x^2(x^2 - 2)$

e $y = \frac{1}{4}x^2(x + 4)^2$ **f** $y = x(x^2 + 1)^2$

8 A curve has equation $y = (x - 2)(x^2 + 2x + 4)$.

a Express the equation of this curve in the form $y = x^3 + b$ for b a constant.

b Find the gradient of this curve at the point where

i $x = \frac{1}{3}$ **ii** $y = 0$

9 A curve has equation $y = 4x^2 - 3x + 2$.

a Find $\frac{dy}{dx}$.

b Find the value of x for which $\frac{dy}{dx} = 0$.

The gradient of the curve at a particular point P is 1.

c Find the coordinates of P.

10 a Find the coordinates of the point on the curve $y = 3x^2 - 2x - 1$ where the gradient is 10.

b Find the coordinates of the point on the curve $y = \frac{1}{4}x(2x + 3)$ where the gradient is $\frac{7}{4}$.

c Find the coordinates of the point on the curve $y = \frac{x}{2}(x^3 + 24)$ where the gradient is -4.

11 The diagram shows the graph of $y = x^3 + 3x^2 + 2$. Also shown are the tangents to this curve at the point $P(1,6)$ and the point Q.

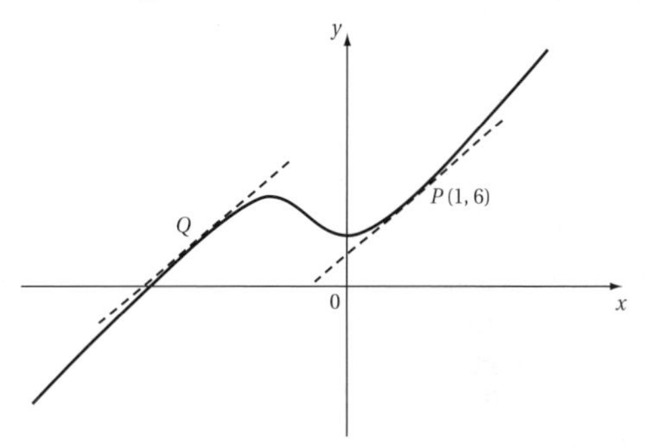

a Show that the gradient of the tangent at point P is 9.

The tangent at Q is parallel to the tangent at P.

b Find the coordinates of point Q.

12 The diagram shows the graph of $y = kx^2 + 3$ where k is a constant, which passes through the points A, B and C. The x-coordinates of A and C are 2 and 3, respectively. The x-coordinate of B is p, where $2 < p < 3$.

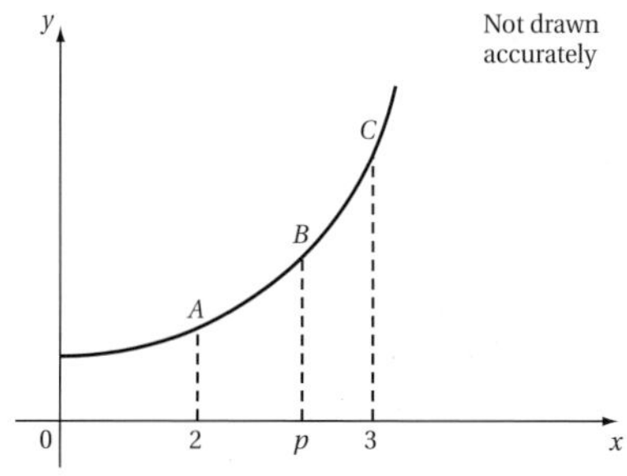

The gradient of this curve at point A is 1.

a Show that $k = \frac{1}{4}$.

b Hence find the gradient of this curve at point C.

Sanjay calculates the gradient of the curve at B. His answer is $\frac{7}{4}$.

c Explain why Sanjay's answer must be wrong.

13 By expressing these equations in an appropriate form, find $\frac{dy}{dx}$.

a $y = \frac{x^4 + x}{x}$ **b** $y = \frac{2x^5 - 3x^3 + x^2}{x^2}$

c $y = \frac{4x^3 - 9x^2}{2x}$ **d** $y = \frac{x^2 - 4}{x + 2}$

e $y = \frac{2x^2 + x - 1}{3x + 3}$ **f** $y = \frac{x^4 - 1}{x^2 + 1}$

14 In this question you should use differentiation from first principles.

A curve has equation $y = \frac{1}{x}$ where $x \neq 0$. $Q\left(q, \frac{1}{q}\right)$ is a point on this curve which is close to the general point $P\left(x, \frac{1}{x}\right)$.

a Show that $Grad_{PQ} = -\frac{1}{qx}$.

b Hence find an expression for $\frac{dy}{dx}$ in terms of x for this curve.

c Verify that your answer to part **b** agrees with that found by using the rule 'If $y = x^n$, where n is any number, then $\frac{dy}{dx} = nx^{n-1}$'.

15 In this question you should use differentiation from first principles.

A curve has equation $y = \sqrt{x}$ where $x \geqslant 0$. $Q(q, \sqrt{q})$ is a point on this curve which is close to the general point $P(x, \sqrt{x})$.

a Expand and then simplify $(\sqrt{q} + \sqrt{x})(\sqrt{q} - \sqrt{x})$.

b Hence show that $Grad_{PQ} = \frac{1}{\sqrt{q} + \sqrt{x}}$.

c Find an expression for $\frac{dy}{dx}$ in terms of x for this curve.

d Verify that your answer to part **c** agrees with that found by using the rule 'If $y = x^n$, where n is any number, then $\frac{dy}{dx} = nx^{n-1}$'.

8.3 Applications of differentiation

1 The diagram shows the curve $y = 2x^2 - 5x + 6$. The tangent T to this curve at point $P(2,4)$ has also been drawn.

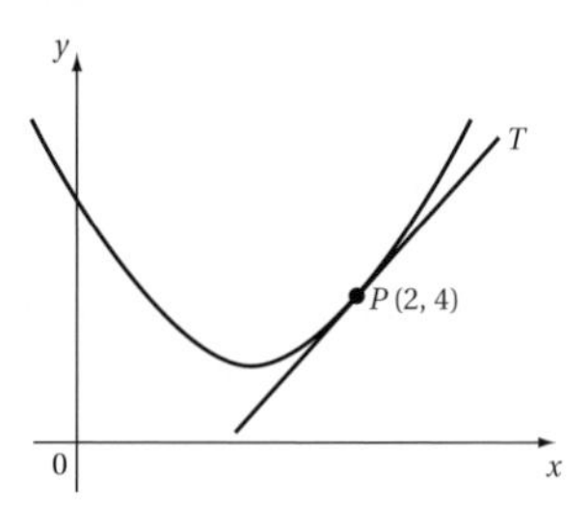

a Show that the gradient of T is 3.

b Hence find an equation for T.

2 Find the equation of the tangent to these curves at the given point P. Give each answer in the form $y = mx + c$ where m and c are constants.

a $y = 3x^2 + 2x + 5$ at the point $P(-1,6)$

b $y = (2x - 1)(x - 1)$ at the point $P(3,10)$

c $y = x^3 - 2x^2 + 3x - 1$ at the point $P(1,1)$

d $y = 4 + 3x - 4x^2$ at the point where $x = 1$

3 Find the equation of the tangent to the curve $y = (x^2 - 2)(x - 4)$ at the point where $x = 2$.

4 A curve has equation $y = 4x^2 - 6x + 1$.

a Show that an equation for the tangent to this curve at the point where $x = \frac{1}{2}$ is $y + 2x = 0$.

b Find an equation for the tangent to this curve at the point where $x = \frac{1}{4}$. Give your answer in the form $ay + bx = c$ for integers a, b and c.

5 The diagram shows the curve $y = x^2 - 4x + 7$ which passes through the point $P(3,4)$. The normal N to this curve at the point P is also shown.

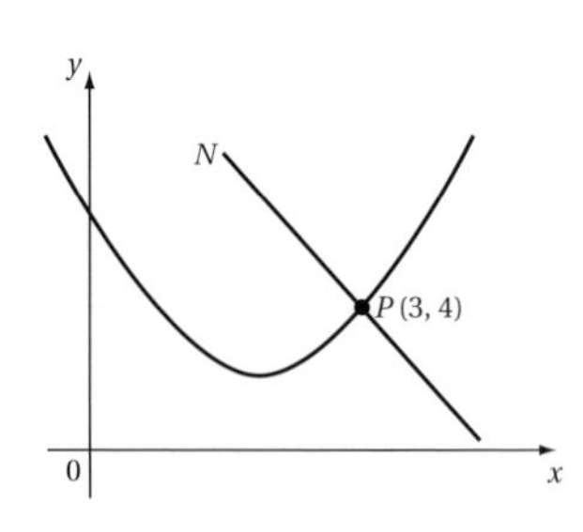

a Find the gradient of the tangent T to the curve at P.

b Hence show that an equation for N is $2y + x = 11$.

6 Find an equation for the normal to these curves at the given point. Give each answer in the form $ay + bx + c = 0$ for integers a, b and c.

a $y = 9 - x - x^2$ at the point $P(2,3)$.

b $y = (x^2 - 3)(2x + 1)$ at the point $P(-1,2)$.

c $y = (x^2 + 1)(x^2 - 8)$ at the point where $x = 2$

7 The diagram shows the curve $y = 3x^2 - 5x + 7$. The curve crosses the y-axis at point A. The tangent T to this curve at point P is also shown. The x-coordinate of P is 2 and the tangent crosses the y-axis at point Q.

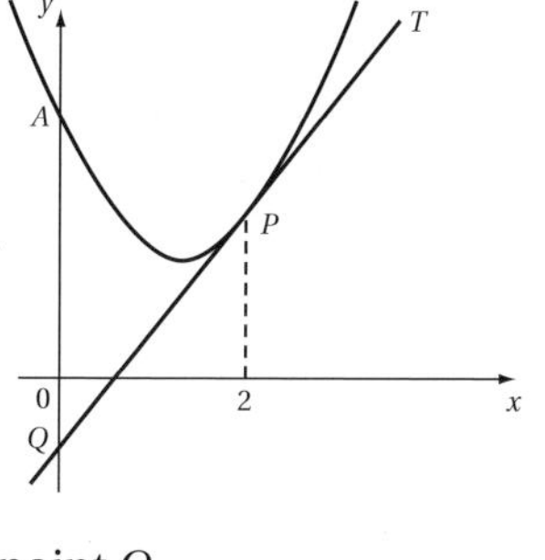

a Write down the coordinates of point A.

b Show that an equation for T is $y = 7x - 5$.

c Find the area of triangle PAQ.

8 The diagram shows the curve $y = 3 + 4x - 4x^2$ which passes through the point P(1,3). The tangent T and the normal N to this curve at point P are also shown. Points A and B are where the tangent T and normal N intersect the y-axis respectively.

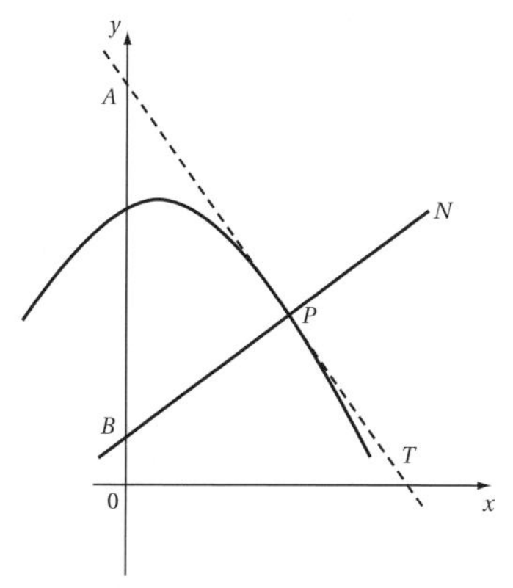

a Show that an equation for T is $y = 7 - 4x$.

b Find an equation for N.

c Show that the length $AB = \frac{17}{4}$.

d Hence, or otherwise,

i find the area of triangle APB

ii find the product of the length AP with the length BP.

9 A curve has equation $y = x^2 + kx - 12$ where k is a constant. The line $y = 2x + 3$ is a tangent to this curve at the point where $x = 3$.

a Show that $k = -4$.

b Hence find an equation for the tangent to this curve at the point where $x = -3$.

c Find the coordinates of the point on this curve where the gradient of the curve is 0.

10 The diagram shows the curve $y = x^2 - 5x + 10$. Also shown is the normal N to the curve at the point P. The x-coordinate of P is 3. N intersects the curve again at point Q.

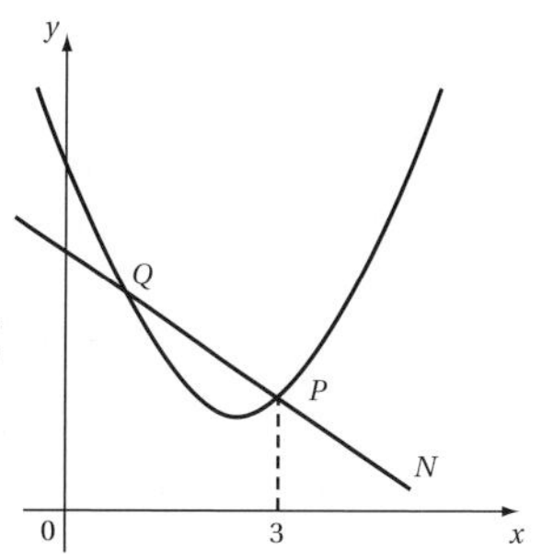

a Show that an equation for N is $y = 7 - x$.

b Use algebra to find the coordinates of point Q.

11 a Find the coordinates of the stationary points on these curves.

i $y = 2x^2 - 12x + 13$

ii $y = 5 - 4x - 4x^2$

b Hence sketch, on separate diagrams, each curve in part **a**. On each sketch, label the y-intercept and the stationary point with their coordinates.

12 The diagram shows the curve $y = 2x^3 + 3x^2 - 12x + 5$. The curve has stationary points at P and Q.

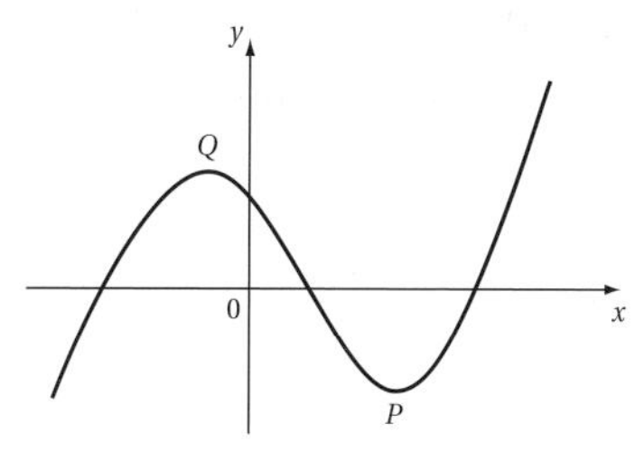

a Show that the x-coordinate of any stationary point on this curve satisfies the equation $x^2 + x - 2 = 0$.

b Hence find the coordinates of P and the coordinates of Q.

13 a Find the coordinates of any stationary points on the curves with these equations.

i $y = x^3 - 12x + 10$

ii $y = x^3 - 6x^2 + 9x - 2$

iii $y = 3x^2 + 7 - x^3$

iv $y = x^2(2x - 9) + 3$

v $y = (x + 1)^2(2 - x) + 1$

vi $y = 4x^3 - 9x^2 + 6x - 4$

b Hence sketch, on separate diagrams, each curve in part **a**. You do not need to find the coordinates of the points where each curve crosses the x-axis.

Handy hint

Use the y-intercept of each curve to help you complete the sketch.

14 Prove that the curves with these equations do not have any stationary points.

a $y = x^3 + x + 1$

b $y = x^3 + 3x^2 + 6x - 1$

c $y = x^3 + kx^2 + k^2x + 1$ where k is a non-zero constant.

15 James is studying AS Maths and says 'Every cubic graph, if drawn for all values of x, must cross the x-axis at least once'.

a Show, with the aid of various sketches, that James is correct.

A quartic graph is a graph with equation $y = ax^4 + bx^3 + cx^2 + dx + e$, where a, b, c, d and e are real numbers.

In response, Alice says 'Any quartic graph must have at least one stationary point'.

b Explain why Alice is also correct.

Tom joins the conversation and says 'Every cubic graph must have a stationary point'.

c State whether or not Tom is correct, justifying your answer.

1 Answers: Numbers and indices

1.1 Fractions

1 **a** $\frac{5}{9}$ **b** $\frac{15}{16}$ **c** $\frac{1}{6}$ **d** $\frac{4}{15}$

2 **a** $\frac{11}{15}$ **b** $\frac{1}{2}$ **c** $\frac{11}{20}$ **d** $\frac{9}{10}$

3 **a** $\frac{8}{3}$ **b** $\frac{23}{5}$ **c** $\frac{62}{11}$ **d** $\frac{25}{9}$

4 **a** $\frac{35}{2}$ **b** $-\frac{4}{3}$ **c** $-\frac{5}{4}$ **d** $\frac{15}{8}$

5 **a** $\frac{12}{5}$ **b** $\frac{32}{5}$ **c** $\frac{15}{8}$ **d** $\frac{8}{15}$

6 **a** $\frac{11}{3}$ **b** $\frac{11}{6}$ **c** $\frac{23}{6}$ **d** $\frac{1}{9}$ **e** $\frac{17}{20}$ **f** $\frac{4}{5}$

7 **a** $\frac{1}{3}$ **b** 4

8 **a** 59°F **b** $18\frac{1}{2}$ °F (or $\frac{37}{2}$ °F)

9 **a** $5\frac{5}{6}$ miles (or $\frac{35}{6}$ miles) **b** 6 km

10 **a** $\frac{8}{3}$ **b** $-\frac{3}{10}$ **c** $-\frac{7}{11}$ **d** $\frac{4}{5}$

11 **a** Hint: Solve $4\frac{2}{5} \times w = 11$ by converting $4\frac{2}{5}$ to a top-heavy fraction.

b $13\frac{4}{5}$ m

1.2 Surds

1 **a** $7\sqrt{5}$ **b** $6\sqrt{6}$ **c** $2\sqrt{2}$

d 20 **e** $-16\sqrt{2}$ **f** $2\sqrt{7}$

2 **a** $5\sqrt{5}$ **b** $\sqrt{2}$ **c** $-4\sqrt{3}$

3 3

4 **a** $4 + 3\sqrt{2}$ **b** $1 - 3\sqrt{3}$ **c** $1 + 5\sqrt{2}$

d $10 + 4\sqrt{6}$ **e** $2 + 7\sqrt{3}$ **f** $5 + 2\sqrt{6}$

5 **a** $2 - \sqrt{3}$ **b** $6 + 2\sqrt{3}$ **c** $6 - 2\sqrt{3}$

6 **a** $11 + 4\sqrt{7}$ **b** $5 + 2\sqrt{3}$ **c** $-5 - 3\sqrt{3}$

7 -2

8 **a** **i** $2\sqrt{2}$ **ii** $2\sqrt{10}$

b Because $D = -3b^2 < 0$ for any $b \neq 0$ and the square root of a negative number does not have a real value.

9 **a** $5\sqrt{3}$ **b** $6\sqrt{5}$ **c** $8\sqrt{3}$ **d** $9\sqrt{6}$

10 **a** 2

b Hint: Show $(4 \pm 2\sqrt{3})^2 = 28 \pm 16\sqrt{3}$ and then use Pythagoras' theorem.

c $8 + 2\sqrt{14}$, $a = 8$, $b = 2$

11 **a** $2 + 2\sqrt{3}$

b Hint: Show the three numbers obey Pythagoras' theorem $c^2 = a^2 + b^2$, where $c = 2 + 2\sqrt{3}$

c $3 + 2\sqrt{3}$, $a = 3$, $b = 2$

12 **a** $6 + 2\sqrt{5}$

b Hint: Use Pythagoras' theorem to find a shorter side.

c $2 + 2\sqrt{5}$

d Both equal $10 + 6\sqrt{5}$

1.3 Indices

1 **a** 2^7 **b** 2^9 **c** 2^{10} **d** 2^{24}

2 **a** $\frac{1}{16}$ **b** $\frac{1}{16}$ **c** $\frac{1}{125}$ **d** $\frac{1}{54}$

3 **a** 3 **b** 6 **c** 2 **d** 2

4 **a** 8 **b** 9 **c** 32 **d** 27 **e** $\frac{1}{4}$ **f** $\frac{1}{2}$

g $\frac{1}{25}$ **h** $\frac{1}{256}$

5 $\frac{5}{4}$

6 **a** $\frac{4}{3}$ **b** $\frac{3}{2}$ **c** $\frac{2}{3}$ **d** 7

7 a $2x^{-1}$ **b** $\frac{1}{2}x^{-3}$ **c** $3x^{\frac{3}{2}}$

d $\frac{1}{4}x^{\frac{2}{3}}$ **e** $2x^{-\frac{1}{2}}$ **f** $3x^{\frac{2}{3}}$

g $\frac{1}{2}x^{-\frac{1}{2}}$ **h** $10x^{\frac{1}{4}}$

8 a True (add indices)

b False (e.g. $n = 2, a = 3, b = 2$)

c False (e.g. $a = 2, m = 2, n = 1$)

d True (add indices, $a^0 = 1$)

e False (e.g. $a = 2, n = 3$)

9 a 24 **b** $\frac{8}{3}$ **c** $\frac{5}{4}$

d 0 **e** 4 **f** $\frac{35}{6}$

10 a $3x + 2x^{-2}$ **b** $1 - \frac{3}{2}x^{-1} + \frac{1}{2}x^{-2}$

11 a $2x - 1 - x^{-1}$ **b** $9x^{-1} + 12x^{-2} + 4x^{-3}$

c $x^{\frac{3}{2}} + 3x^{\frac{1}{2}} - 6x^{-\frac{1}{2}}$ **d** $4x^{-2} + 4x^{-\frac{3}{2}} + x^{-1}$

12 a $6 + 13x^{-1} + 6x^{-2}$

b $13x^{-1} = \frac{13}{x}$ and $6x^{-2} = \frac{6}{x^2}$ so as x increases, these fractions approach 0.
So $6 + 13x^{-1} + 6x^{-2}$ approaches 6.

c Yes: the point $(-\frac{6}{13}, 6)$

2 Answers: Algebra 1

2.1 Basic algebra

1 a $5a + 3$ **b** $-5b + 18$ **c** $10a$ **d** $2a^2 + 3b^2$

2 a $xy(2x + y)$ **b** $2x^2y^2(5x - 2y)$

c $3x^3yz(xy + 2z)$ **d** $3xy(4x^3y + 2xy - 3)$

3 a $4a^4b^2$ **b** $9a^3b^2(3b^4 + a)$

c $4a^2b^4(2a + b)(2a - b)$

4 a $Q = \frac{P}{3} - 4$ (or $\frac{P - 12}{3}$)

b $B = \frac{2A + 1}{3}$ **c** $T = \frac{R + 3}{2}$

d $D = \frac{2C - 5}{12}$ **e** $V = 9U^2 - 2$

f $N = \sqrt[3]{\frac{2M}{\pi}} + 1$

5 a $r = \sqrt[3]{\frac{3V}{4\pi}}$ **b** 3 cm

6 a Hint: The quarter-circle has area $\frac{1}{4}\pi(2x)^2$

b $x = \sqrt{\frac{A}{4 - \pi}}$

c Hint: The quarter-circle has perimeter $\frac{1}{4}(2\pi)(2x)$

7 a Hint: Start by expressing t in terms of m using $\tan \hat{A} = \frac{\text{opposite}}{\text{adjacent}}$

b $m = \sqrt{\frac{n^2 - 2}{2}}$

8 a $x = \pm\sqrt{y} - 3$

b $x = 1 \pm \sqrt{\frac{y + 1}{4}}$ (or $x = 1 \pm \frac{1}{2}\sqrt{y + 1}$)

c $x = \frac{5 \pm \sqrt{3y}}{2}$

9 a Hint: Use the rule $\frac{a + b}{c} = \frac{a}{c} + \frac{b}{c}$

b $Q = \frac{3}{P - 2}$

10 a $B = \frac{2}{1 - A}$ **b** $D = \pm\sqrt{\frac{4}{C - 1}}$

c $F = \sqrt[3]{\frac{5}{E + 4}}$

11 a $B = \frac{2A}{A - 1}$ **b** $D = \frac{2 - 3C}{2C - 1}$ **c** $F = \pm\sqrt{\frac{3 - E}{E - 1}}$

12 a $x + 3$ **b** $2x^2 + 4$ **c** $3x$ **d** $-x^2$

13 a $x^2 + \frac{4}{3}x + 2, A = 1, B = \frac{4}{3}, C = 2$

b $2x - \frac{3}{2}x^{-1}, A = 2, B = \frac{3}{2}$

c $2x^2 + \frac{1}{3}x^{-2} + 3, A = 2, B = \frac{1}{3}, C = 3$

d $4x^3, A = 4, n = 3$

2.2 Solving linear equations

1 a 5 **b** 8 **c** 3 **d** −3

2 a 5 **b** $\frac{1}{5}$ **c** −4

3 a 4 **b** 8 **c** 6

4 a 3 **b** $\frac{1}{2}$ **c** $\frac{1}{4}$

5 a 4 **b** $\frac{3}{2}$

6 a Hint: Substitue $x = 6$ and $y = 3$ into the equation.

b 8

7 a $x = 2, y = 3$ **b** $x = 7, y = -1$

c $x = \frac{1}{2}, y = -\frac{3}{2}$

8 a $2(5x - 1)$ **b** $\frac{7}{2}$ **c** 68 cm^2

9 a Hint: The circumference of a semi-circle radius x is πx.

b 18 800 cm^2 (3 sig. figs)

10 a 7 **b** 12 **c** 3

2.3 Linear inequalities

1 a $x \leqslant 2$ **b** $x > -4$ **c** $x \geqslant -2$ **d** $x \geqslant \frac{3}{5}$

2 a $0 \leqslant 3$ so $x = 0$ satisfies Jane's solution.

$x = 0$: $9 - 4x = 9 - 4(0)$

$= 9$, which is **not** less than -3.

So Jane's solution is incorrect.

b When Jane cancelled off the minus signs she was actually multiplying both sides by −1, and so she should have reversed the inequality symbol.

c $x \geqslant 3$

3 a $x \geqslant 2$ **b** $x > \frac{1}{2}$ **c** $x > 8$

d $x \geqslant -\frac{1}{2}$ **e** $x \leqslant -2$ **f** $x > 1$

4 a $1 \leqslant x \leqslant 4$ **b** $-3 < x \leqslant -1$

c $-3 < x < -2$

5 a $\frac{4}{3} \leqslant x \leqslant \frac{11}{3}$ **b** $x = 2, x = 3$

6 a $0 \leqslant x < \frac{2}{3}$

b Unsolvable, as 12 is not less than 6

c $x = -\frac{3}{4}$

7 a $2 < x \leqslant 4$

b $2x + 3 > 16$ has solution $x > \frac{13}{2} = 6.5$

$4 - 3x \geqslant x - 20$ has solution $x \leqslant 6$

8 $3 < x \leqslant 4$

9 a Since $-7 < -4$, any number less than −7 must also be less than −4. But Tom's expression says $x > -4$.

b $-7 < x < -4$

10 a $x \geqslant 2$ **b** $14x - 10$ **c** 5

11 a $x \geqslant \frac{3}{2}$ **b** $x = \frac{3}{2}$

c The solution to $2x - 3 \geqslant 0$ corresponds to the values of x for which the line lies on or above the x-axis, i.e. $x \geqslant \frac{3}{2}$.

12 a $x \leqslant \frac{3}{4}$ **b** $x > -\frac{2}{3}$ **c** $x \leqslant \frac{5}{4}$

2.4 Forming expressions

1 a $6x + 8$

b Hint: Length $BE = \frac{1}{2}(2x + 4) = x + 2$.

Area of trapezium $= \frac{1}{2}(a + b)h$, where a, b are parallel sides and h is the height.

2 a i $10(x + 1)$ **ii** $5x(x + 3)$ **b** 350 cm^2

3 a $2\pi(x + 3)$

b Hint: The area of larger circle is $\pi(x + 3)^2$.

4 a Hint: The quarter-circle has circumference $\frac{1}{2}\pi r$.

b $\frac{1}{4}r^2(\pi - 2)$

5 a $P = 4x$ **b** $x = \sqrt{2}y, k = \sqrt{2}$

6 a $2x + y = 24$

b Hint: Start by rearranging $2x + y = 24$ to make y the subject.

c 64 m^2

7 a $V = 6x^3$

b Hint: The cuboid has 6 faces. The base has area $2x^2$ etc.

c $\frac{3}{11}$ **d** 88 cm^2

8 a $48 - 4x^2$

b Hint: Label the sides of the tray with its dimensions. e.g. The height of the tray is x cm.

c 1.5 cm **d** 22.5 cm^3

9 a $S = 2xy + 8x + 8y$

b Hint: Use Volume = base × width × height.

c $S = 8\left(x + \frac{4}{x} + 1\right)$

10 a $V = \pi r^2 h$ **b** $r = \frac{3}{\sqrt{h}}$

c Use Pythagoras' theorem where L is the hypotenuse.

11 a $4L$

b By construction, the circular lid has length L and so L = circumference of circle, radius r. So $L = 2\pi r$ (that is, $r = \frac{L}{2\pi}$).

c Hint: Use $V = \pi r^2 h$ where $h = 4$ and $r = \frac{L}{2\pi}$.

d Hint: Add on the two circular parts, each with area πr^2.

3 Answers: Coordinate geometry 1

3.1 Straight-line graphs

1 a

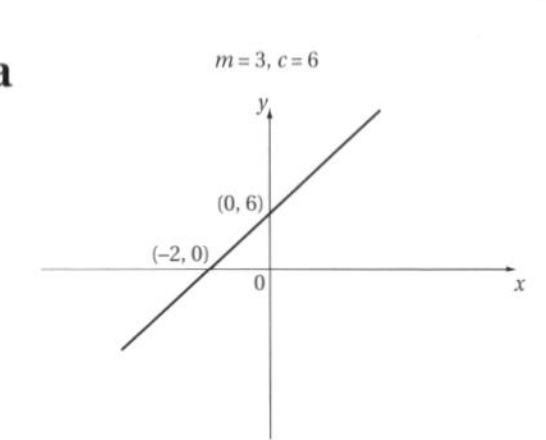

b

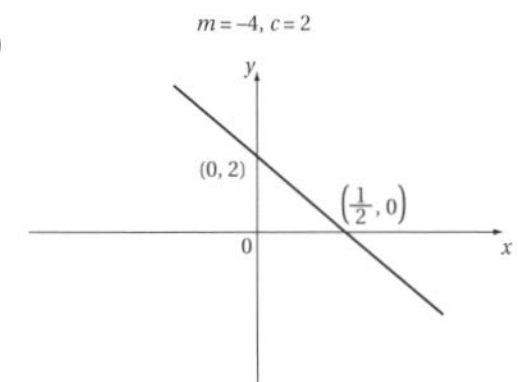

c

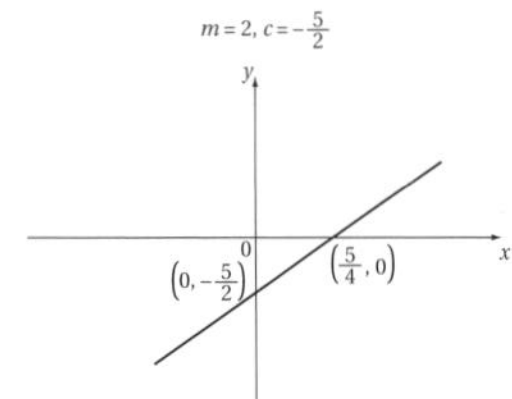

d

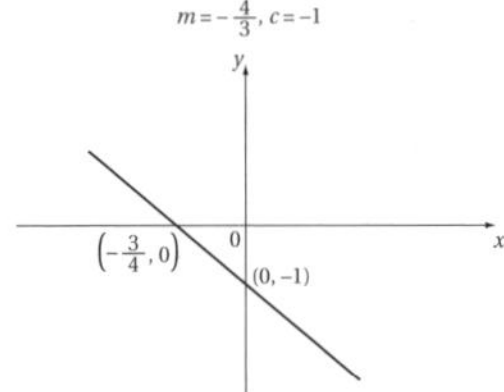

2 a $m = 2, c = -1$ **b** $m = \frac{3}{2}, c = 1$

c $m = \frac{4}{3}, c = -\frac{1}{3}$ **d** $m = -2, c = 12$

3 a

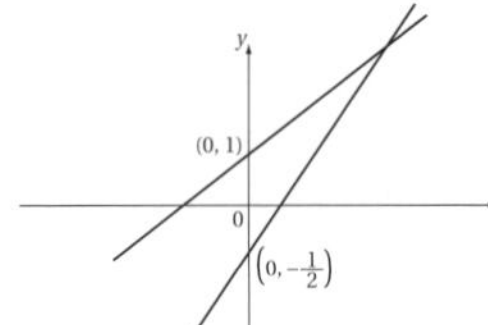

b $\frac{3}{2}$

4 a

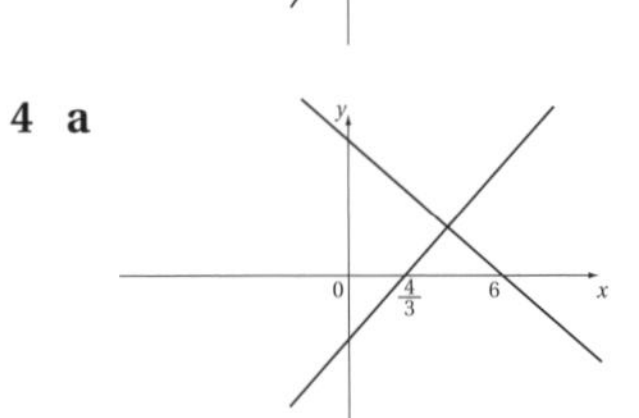

b $\frac{14}{3}$

5 a $2y + x + 3 = 0$ **b** $3y + 2x - 1 = 0$

c $4y + 3x - 2 = 0$ **d** $6y - 4x + 15 = 0$

6 A = (1), B = (4), C = (2), D = (3)

7 a Straight line:

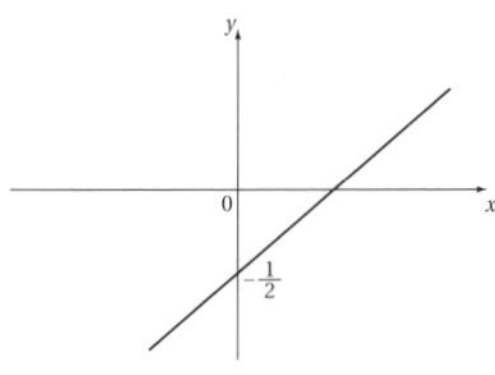

b Straight line:

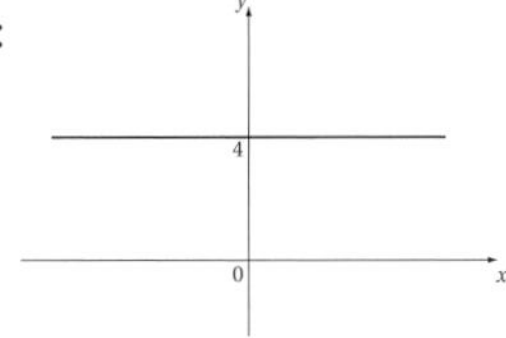

c Not a straight line

d Straight line:

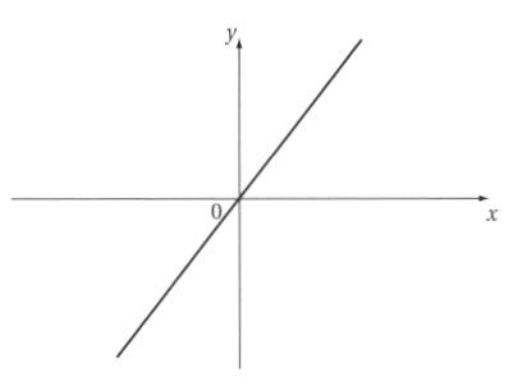

e Not a straight line

f Straight line:

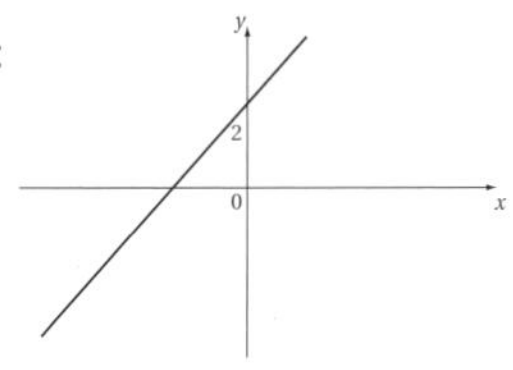

8 a $-\frac{1}{2}$ **b** $-\frac{2}{3}$ **c** $6y + 3x + 4 = 0$

9 a $a = 2, b = -5$ **b** 15

3.2 The equation of a line

1 a 2 **b** 5 **c** $\frac{1}{2}$ **d** -3

2 a $-\frac{1}{2}$ **b** $2y + x = 6$ **c** (6,0)

3 a $y = 4x - 5$ **b** $y = -7$ **c** $2y + 3x = 8$

4 a

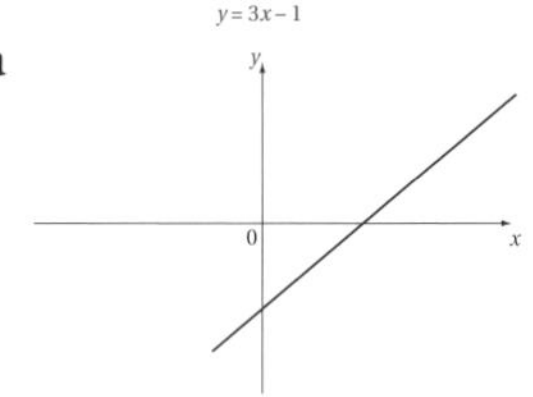

b

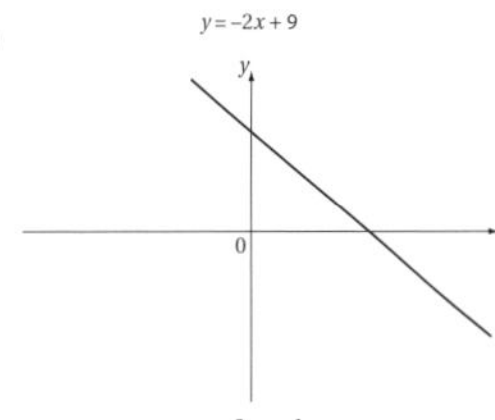

c

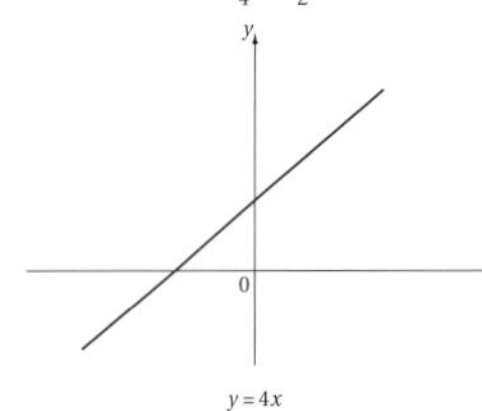

d

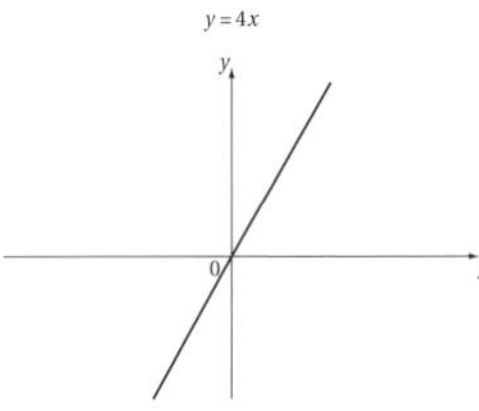

5 a 13 **b** 5 **c** 2

6 a $y = \frac{3}{2}x + 3$ **b** (0,3) **c** $(-2,0)$ **d** 3

7 a $A(0,2), B(\frac{3}{2}, 0)$

b Hint: Use Pythagoras' theorem on the triangle *OAB*, where *O* is the origin.

8 a -2

b *B* does not lie on this line. If you substitute $x = -7$ into the equation, $y = -2(-7) - 3$
$= 14 - 3$
$= 11$

which is not the y-coordinate of *B*.

9 a Hint: Substitute $x = \frac{5}{2}, y = \frac{1}{2}$ into the equation.

b $-\frac{9}{2}$ **c** $\frac{81}{16}$

10 a Hint: Simplify $\frac{4 - 4k}{k - 1}$

b $y = -4x + 4k + 4$

c Hint: Make $y = 0$ in the equation and find x.

d 2

11 a Hint: Start by showing the gradient of the line is $\frac{2}{3}$.

b

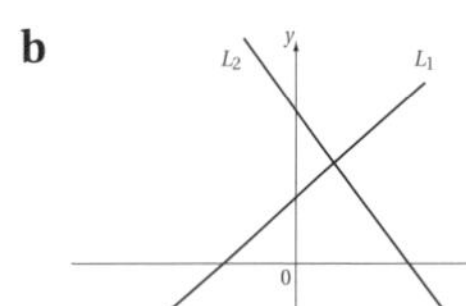

c Hint: Show that the values $x = 6$ and $y = 5$ satisfy both equations

d 39

3.3 Mid-points and distances

1 **a** $(6,4)$ **b** $(2,3)$ **c** $\left(-\frac{3}{2}, \frac{15}{2}\right)$ **d** $\left(2, \frac{11}{12}\right)$

2 **a** $\left(\frac{3+p}{2}, \frac{2+q}{2}\right)$ **b** $p = 5, q = 0$

3 **a** $\left(\frac{k+2}{2}, k+2\right)$ **b** $(-k, k+3)$ **c** $(2k, 2k+1)$

4 $(3,8)$

5 **a** $(-2,10)$ **b** $(-1,13)$

6 **a** Hint: Solve $\frac{p+14}{2} = 8, q = 19$

b $(5,7)$ **c** 15

7 **a** $(1,7)$ **b** $(1,11)$

8 **a** $\sqrt{29}$ **b** $3\sqrt{10}$ **c** $2\sqrt{26}$ **d** $\frac{5}{4}$

9 Hint: Calculate each distance AB, AC and BC. You will see that exactly two of these distances are equal.

10 **a** $\left(6, \frac{1}{2}\right)$ **b** Hint: Find the midpoint of AB.

11 **a** Hint: Find the length AC. **b** $(-3, -15)$

12 **a** $AB = \sqrt{45}, BC = \sqrt{5}, AC = \sqrt{50}$

b Hint: Use the distance formula.

c It is a right-angled triangle (with the right-angle at vertex B).

13 **a** Hint: Solve the equation $\sqrt{(p-1)^2 + 9} = p$

b $\left(3, \frac{9}{2}\right)$ **c** 5π

3.4 Parallel and perpendicular lines

1 **a** -3 **b** $-\frac{1}{4}$

2 **a** $\frac{5}{2}$ **b** $\frac{3}{4}$

3 **a** common gradient 2 **b** common gradient $\frac{3}{2}$

c common gradient $-\frac{1}{2}$

4 **a** $3 \times -\frac{1}{3} = -1$ **b** $-\frac{3}{2} \times \frac{2}{3} = -1$

c $-\frac{5}{3} \times \frac{3}{5} = -1$

5 **a** $y = -4x + 3$ **b** $y = \frac{1}{4}x + 2$

6 **a** $\frac{3}{4}$ **b** $3y + 4x = 6$

7 **a** Hint: Substitute $x = -1, y = k$ into the equation.

b $3y + x = 7$

8 **a** $Grad_{AB} = \frac{5}{4}, Grad_{BC} = -\frac{4}{5}$ **b** $\angle ABC = 90°$

c Hint: Find the lengths AB and BC
$\angle BAC = 45°$

9 **a** -2

b Hint: Find an expression for the gradient of BC in terms of k.

c $8y - 14x + 37 = 0$

10 **a** $-\frac{1}{3}$ **b** $3y + x = 19$ (or any equivalent form)

c Hint: Find the gradient of CM where M is the midpointof AB. You will find a sketch helpful.

d Hint: Find the distance CM.

3.5 Intersections of lines

1 **a** **i**

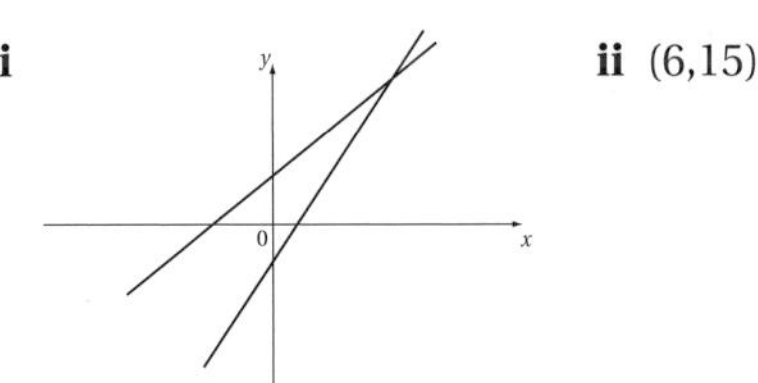

ii $(6,15)$

b **i**

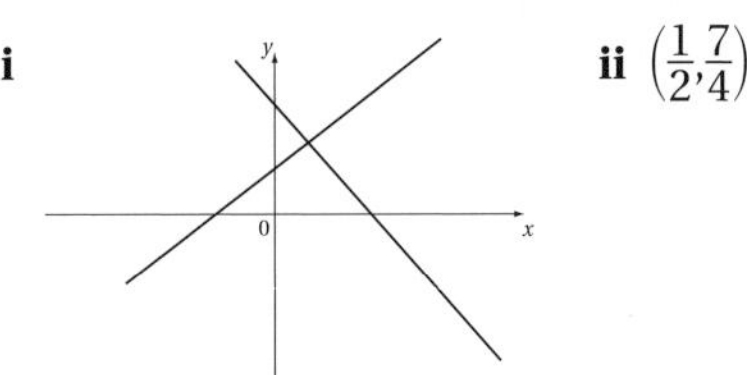

ii $\left(\frac{1}{2}, \frac{7}{4}\right)$

c **i**

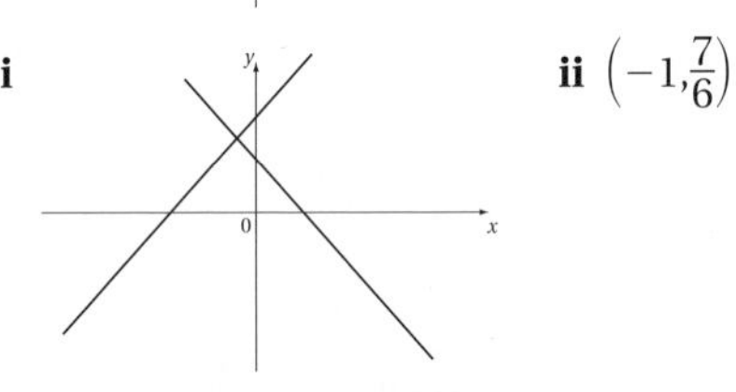

ii $\left(-1, \frac{7}{6}\right)$

2 **a** $(3,-2)$ **b** $(-2,0)$ **c** $(-4,-4)$

3 **a** Hint: Show that trying to solve the equations leads to an impossible result.

b The lines are parallel.

c

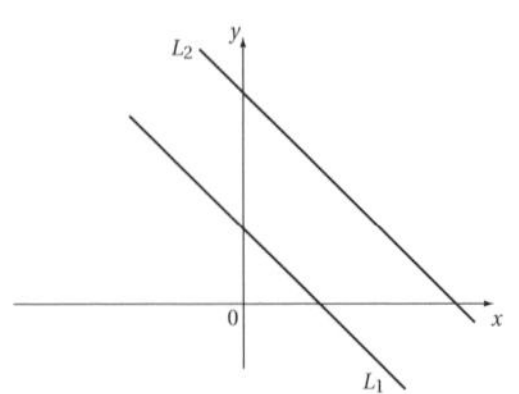

4 **a** Lines intersect at $(3,-2)$.

b Lines do not intersect (parallel, gradient $\frac{3}{5}$).

c Lines intersect at $(0, \frac{1}{3})$.

5 a i

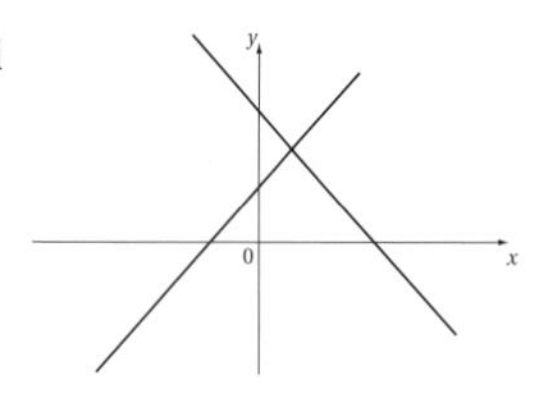

ii $\left(2, \frac{7}{2}\right)$

b i

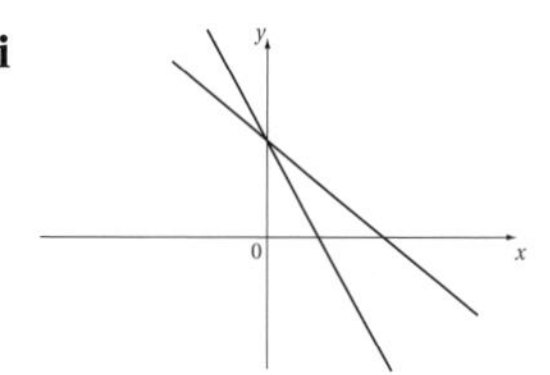

ii (0,4)

c i

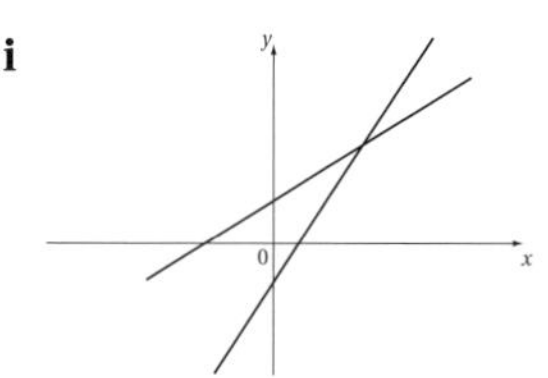

ii (5,3)

6 a $A: x = 1, B: x = \frac{7}{2}$

b Hint: Solve the equations $y = 3x - 3, y = 7 - 2x$ by substitution.

c $\frac{15}{4}$

7 a (2,4) **b i** $\frac{7}{3}$ **ii** 28

8 a $\left(\frac{2}{3}, 1\right)$ **b** $12y + 9x = 18$

9 a Hint: Use $y = mx + c$ where $m = \frac{1}{4}$.

b (2,5) **c** $y = -x + 17, B(6,11)$ **d** $2\sqrt{13}$

4 Answers: Algebra 2

4.1 Solving a quadratic equation by factorising

1 a $x = 3, x = 5$ **b** $x = 2, x = -7$

c $x = 3$ **d** $x = \pm 3$

2 a $x = 0, x = -5$ **b** $x = 0, x = 6$

c $x = 0, x = \frac{3}{2}$

3 a $a = -3, a = -4$ **b** $b = -2, b = 11$ **c** $c = 2$

4 a $x = 5, x = -\frac{1}{2}$ **b** $x = 1, x = \frac{2}{3}$

c $x = \frac{1}{2}, x = -4$ **d** $x = -2, x = -5$

e $x = 2, x = -\frac{1}{2}$ **f** $x = -2, x = -6$

5 a $p = 4, p = -14$ **b** $n = 4, n = 18$

c $t = 3, t = -\frac{15}{2}$

6 a Hint: Substitute $x = 2$ into the equation. **b** 13

7 $\frac{7}{2}$

8 a $y^2 - 5y + 4 = 0$ **b** $y = 4, y = 1 : x = \pm 2, x = \pm 1$

9 a $x = 1, x = 9$ **b** $x = -8, x = 27$

c $x = 0, x = 3$ **d** $x = 0, x = 2$

10 a Hint: Use Area $= \frac{1}{2}(a + b)h$ where a and b are the lengths of the parallel sides and h is the height of the trapezium.

b Hint: Re-arrange $\frac{3}{2}x^2 + 6x = 48$ by first multiplying all terms by 2.

c 16 cm^2

11 a Hint: Use Pythagoras' theorem to find BC^2 and factorise the answer.

b 336 cm^2

12 a $x = -2, x = 3, x = -4$

b $x = 4, x = -1, x = \frac{3}{2}$

c $x = 0, x = -3, x = 2$

d $x = \frac{2}{3}, x = -2, x = 5$

e $x = -1, x = 2, x = 3$

f $x = \frac{1}{2}, x = 3, x = -3$

g $x = 0, x = \frac{3}{4}$ **h** $x = 0, x = 1, x = -4$

4.2 Completing the square

1 a $(x + 3)^2 + 1$ **b** $(x - 2)^2 - 5$ **c** $(x + 4)^2$

2 a $(x - 4)^2 - 9$ **b** $x = 1, x = 7$

3 a $x = -1, x = -5$ **b** $x = 1 \pm \sqrt{3}$

c $x = 5 \pm 2\sqrt{5}$ **d** $x = \frac{1}{2}, x = \frac{5}{2}$

e $x = \frac{1 \pm \sqrt{3}}{2}$ **f** $x = \frac{-5 \pm 3\sqrt{3}}{2}$

4 a $3(x + 2)^2 - 12, p = 2, q = 12$

b $3(x + 2)^2 - 7$

5 a $2(x + 3)^2 - 1$ **b** $3(x - 3)^2 + 4$

c $4\left(x - \frac{1}{2}\right)^2 + 2$

6 a 1 **b** 5

7 a -51 **b** max $= 51, x = 6$ **c** $k = -36$

8 a $(x - 1)^2 + 2$

b Hint: $(x - 1)^2 + 2 = 0$ leads to $x - 1 = \pm\sqrt{-2}$, where $\sqrt{-2}$ is not a real number.

9 a $(x - 1)^2 + 5 = 0$: No real solutions.

b $(x - 4)^2 - 1 = 0$: Real solutions.

c $(x + 5)^2 = 0$: Real solutions.

d $(x + 7)^2 + 1 = 0$: No real solutions.

e $(x + 2)^2 + k^2 = 0$: No real solutions.

f $(x + k)^2 - 2k^2 = 0$: Real solutions.

10 a $x = 1, x = 4$

b Hint: Rearrange the equation $f(x) = 6x - 14$ as $x^2 - 8x + 18 = 0$ and complete the square.

11 a $(x + 4)^2 - 16 + c$

b Hint: There are no real solutions to this equation when $-16 + c > 0$.

12 $k \leqslant 9$

4.3 Solving a quadratic equation using the formula

1 a $x = -3 \pm 2\sqrt{2}$ **b** $x = -2 \pm \sqrt{7}$

c $x = 3 \pm \sqrt{6}$ **d** $x = \frac{-1 \pm \sqrt{17}}{4}$

e $x = \frac{2 \pm \sqrt{10}}{3}$ **f** $x = \frac{5 \pm \sqrt{31}}{2}$

2 a $x = 1 \pm \sqrt{7}$ **b** $x = \frac{-2 \pm \sqrt{13}}{3}$

c $x = 5 \pm 2\sqrt{7}$

3 a Hint: Show that $b^2 - 4ac < 0$

b Hint: Show that $b^2 - 4ac < 0$

4 a Hint: Start with $AB = \sqrt{(p - 2)^2 + (p - 4)^2}$.

b $3 + \sqrt{7}$

5 a Hint: the width AB of the rectangle equals the diameter $2r$ of the circle. The circle has area πr^2.

b Hint: Start with the equation $\pi r^2 + 6\pi r = 11\pi$ and simplify.

c $-3 + 2\sqrt{5}$ cm

6 a $x = \frac{3 \pm \sqrt{9 - 4k}}{2}$ **b** $x = \frac{k \pm \sqrt{k^2 - 8}}{2}$

c $x = \frac{-5 \pm \sqrt{25 - 12k}}{2k}$

7 a Hint: $b^2 - 4ac = (-2)^2 - 4(1)(p) = 4 - 4p$

b $p > 1$

8 a $a > \frac{4}{3}$ **b** $a = \frac{4}{3}; x = \frac{3}{2}$

9 a Hint: $b^2 - 4ac = (-2k)^2 - 4(1)(-1) = 4k^2 + 4$.

b $k^2 + 1 > 0$ for any value of k, so the formula will produce two distinct solutions.

c $2k$ **d** -1

10 a Hint: Use Pythagoras' theorem to give $x^2 + \frac{1}{x^2} = 14$ and then multiply through by x^2.

b $x^2 = 7 + 4\sqrt{3}$ (ignore other solution as $x > 1$)

c 2 **d** $2 + \sqrt{3}$

e Hint: Start by rationalising $\frac{1}{2 + \sqrt{3}}$.

4.4 Solving linear and non-linear equations simultaneously

1 a $x = -1, y = 2 : x = -2, y = 1$

b $x = -\frac{9}{5}, y = -\frac{13}{5} : x = 1, y = 3$

c $x = -\frac{1}{11}, y = \frac{19}{11} : x = -1, y = -1$

2 a $A\left(\frac{2}{5}, \frac{11}{5}\right), B(2, -1)$

b Hint: Simplify $\sqrt{\left(2 - \frac{2}{5}\right)^2 + \left(-1 - \frac{11}{5}\right)^2}$

3 a $x = -\frac{4}{3}, y = -\frac{11}{3} : x = 2, y = 3$

b $x = 2, y = \frac{1}{2} : x = -1, y = -1$

c $x = \frac{17}{8}, y = -\frac{7}{4} : x = -2, y = 1$

4 a (0,10) **b** (−4,2) **c** 20

5 a (3,1)

b The line is a tangent to the circle at the point (3,1)

6 a Hint: Show that trying to solve the equations leads to a quadratic equation with a negative discriminant.

b The line does not intersect or touch the circle.

7 a $x^2 + y^2 = 5$

b Hint: Show that an expression for the perimeter is $4x + 2y$.

c Larger square: side length = 2 m. Smaller square: side length = 1 m.

8 a $x^2 + y^2 = 36$

b $AB = 4 + \sqrt{2}$ cm, $AC = 4 - \sqrt{2}$ cm

c 7 cm^2

5 Answers: Coordinate geometry 2

5.1 Transformations of graphs

1 a

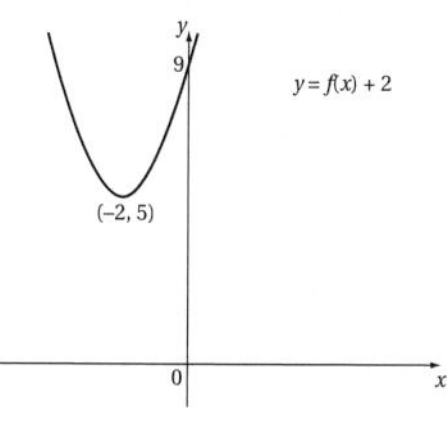

b

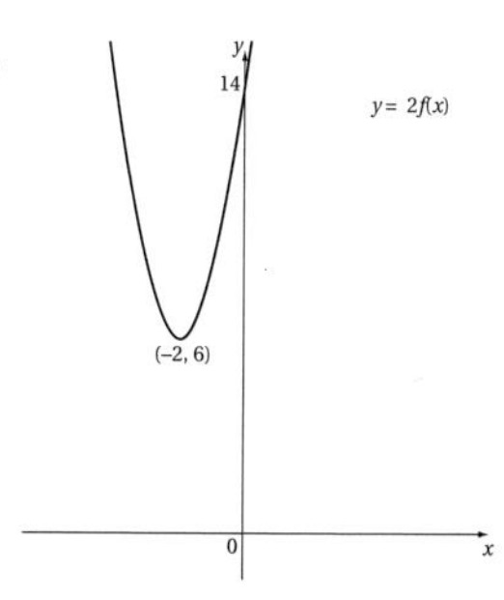

c

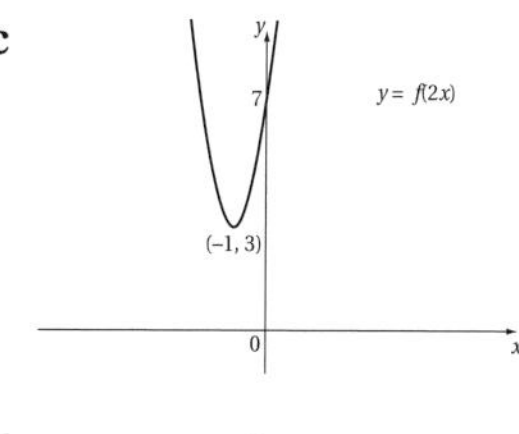

d

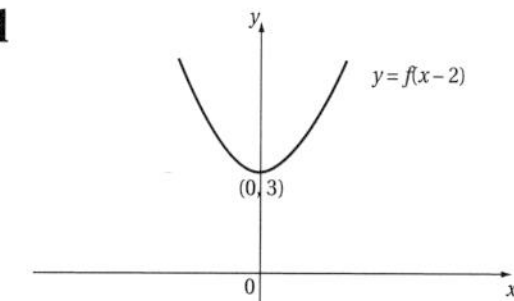

2 a i (4,0) **ii** (3,2) **b**

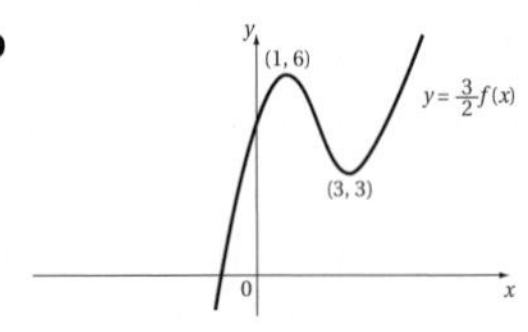

3 a Stretch, scale factor $\frac{1}{a}$ along the x-axis from the origin.

b 4 **c** $\left(\frac{3}{2},0\right), \left(-\frac{1}{4},0\right)$

d i

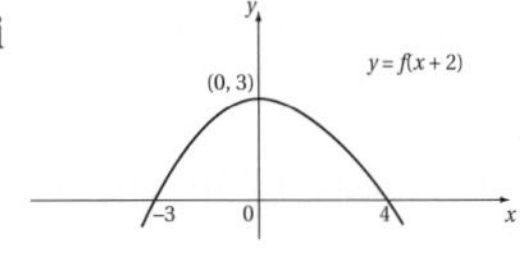

ii

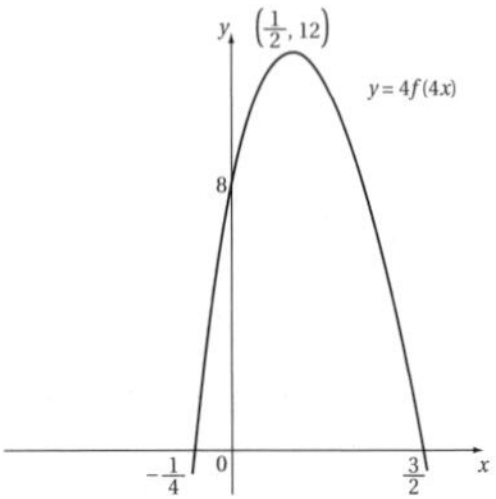

4 a i

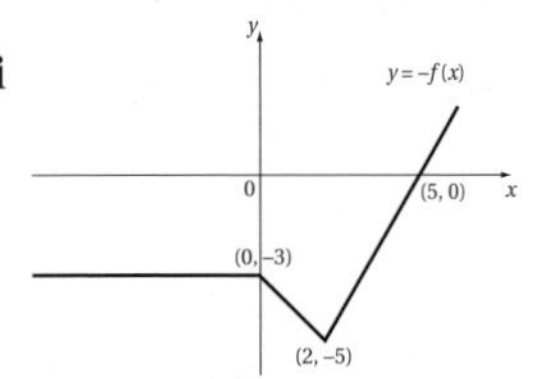

ii

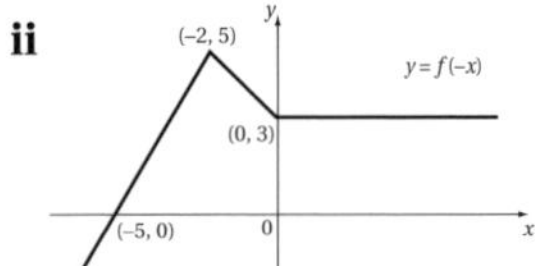

b

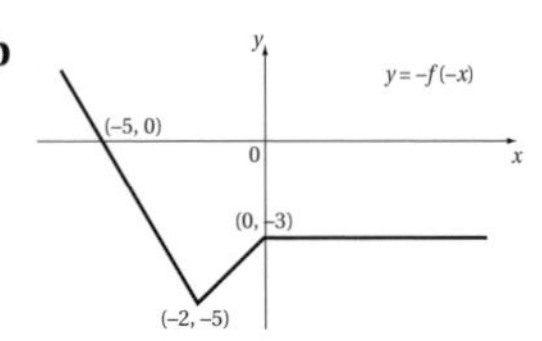

c Rotation 180° about the origin.

5 a i

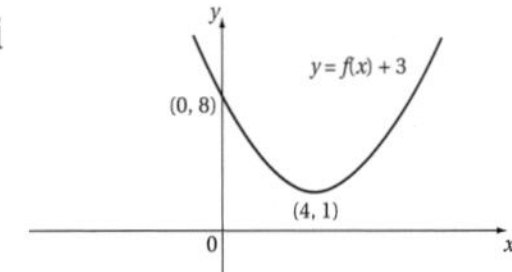

ii

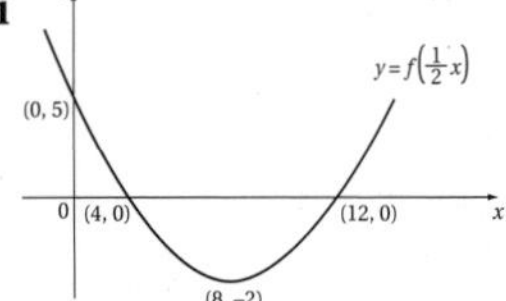

iii

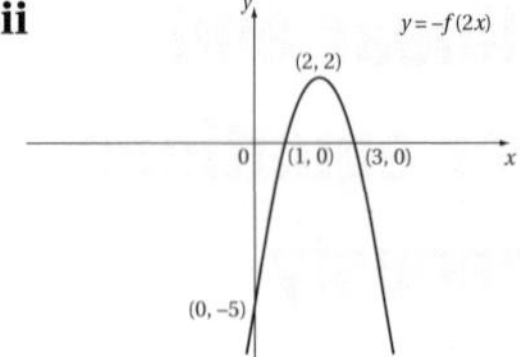

iv

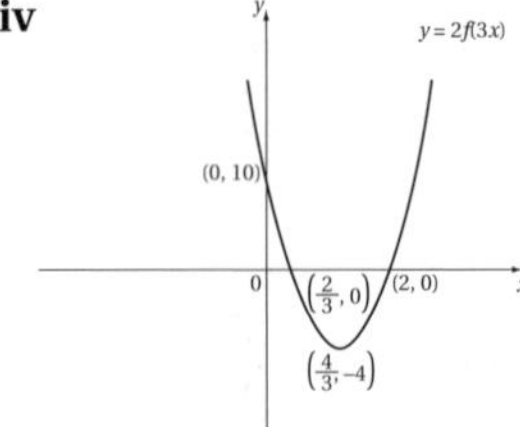

b 2 **c** (2,0), (0,2)

5.2 Sketching curves

1 a i, ii

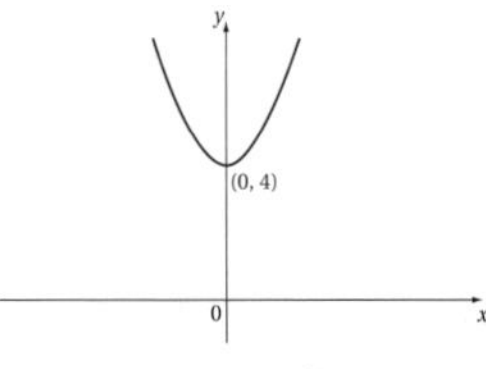

iii Translation by $\begin{pmatrix} 0 \\ 4 \end{pmatrix}$

b i, ii

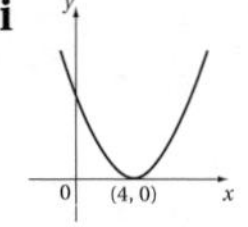

iii Translation by $\begin{pmatrix} 4 \\ 0 \end{pmatrix}$

c i, ii

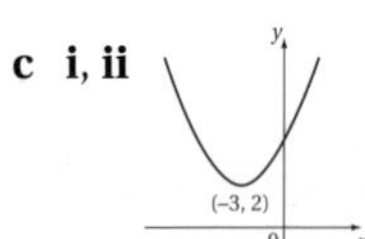

iii Translation by $\begin{pmatrix} -3 \\ 2 \end{pmatrix}$

2 a $(x + 4)^2 - 4$

b

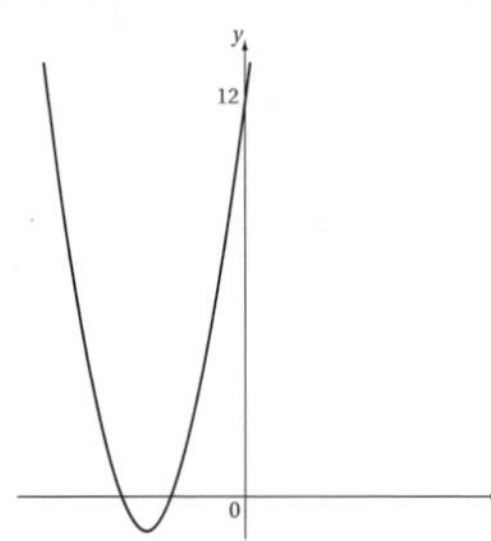

c $x = -4$

3 a $y = (x + 1)^2 + 4$:

Note: Translation of $y = x^2$ through $\begin{pmatrix} -1 \\ 4 \end{pmatrix}$

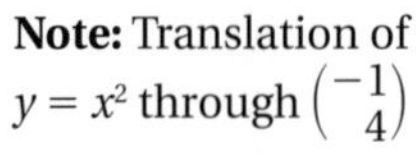

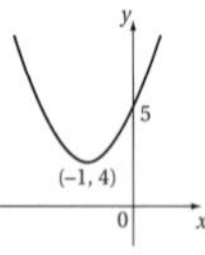

b $y = (x - 3)^2 - 2$:

Note: Translation of $y = x^2$ through $\begin{pmatrix} 3 \\ -2 \end{pmatrix}$

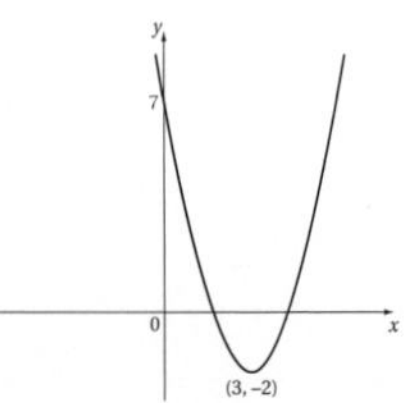

c $y = \left(x + \frac{3}{2}\right)^2 - \frac{13}{4}$:

Note: Translation of $y = x^2$ through $\begin{pmatrix} -\frac{3}{2} \\ -\frac{13}{4} \end{pmatrix}$.

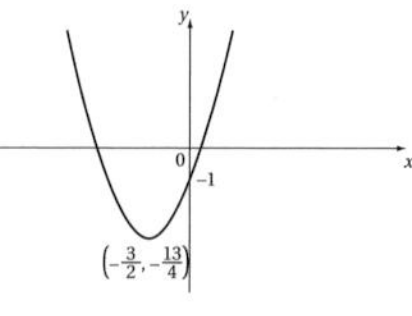

4 a A reflection in the x-axis followed by a translation by $\begin{pmatrix} 0 \\ 1 \end{pmatrix}$.

b

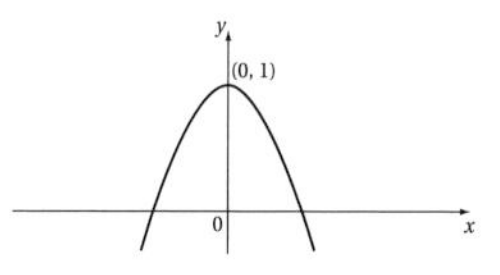

5 a

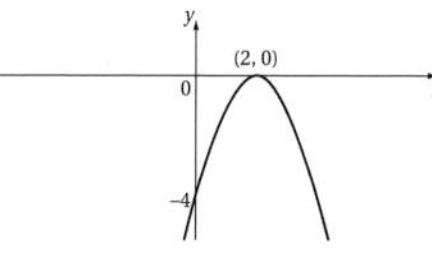

b

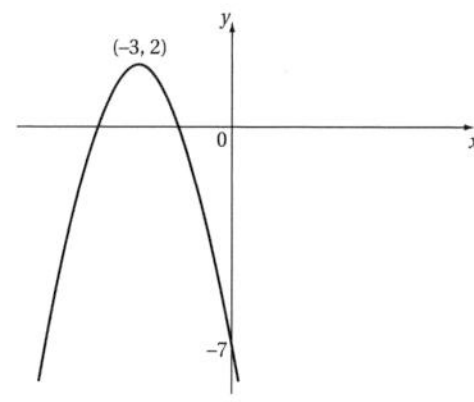

c

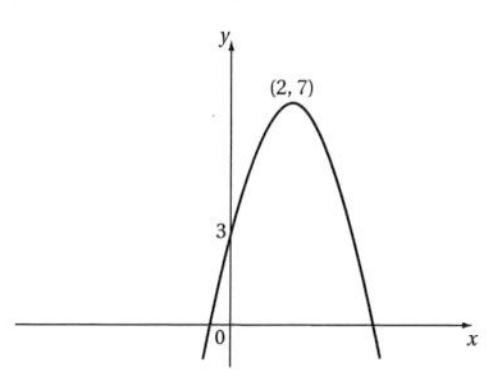

6 a $y = (x - 3)^2 - 4$, $p = 3$, $q = 4$

b Hint: Expand the brackets in $y = (x - 3)^2 - 4$.

c 5

7 a $x = 4$ **b** -4 **c** -9

d Translation by $\begin{pmatrix} 4 \\ -9 \end{pmatrix}$ **e** (0,7)

8 In each answer, the dotted graph is the curve $y = x^2$.

a

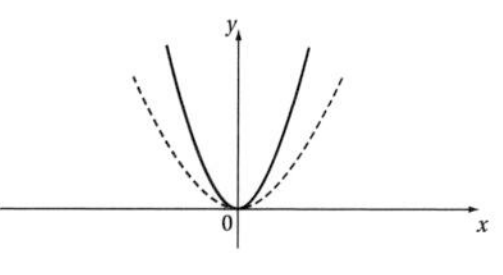

b

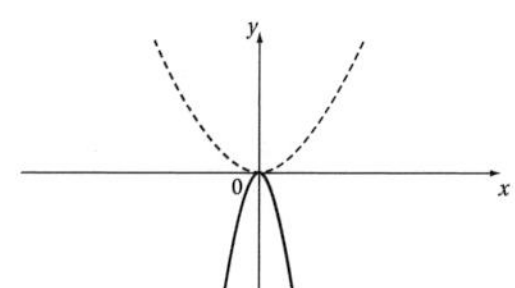

c

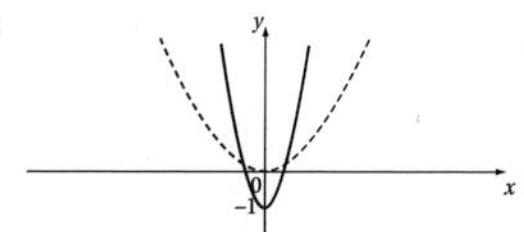

d

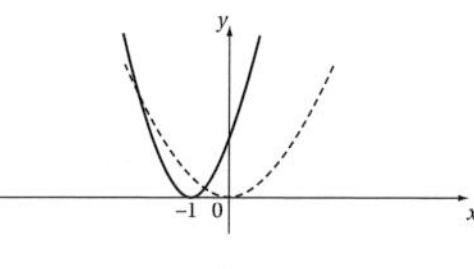

e

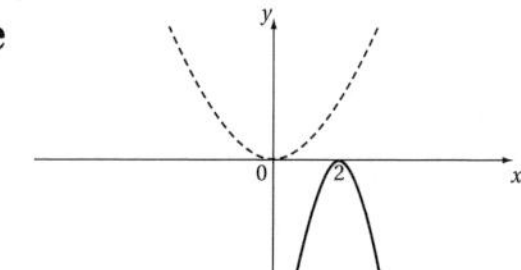

f

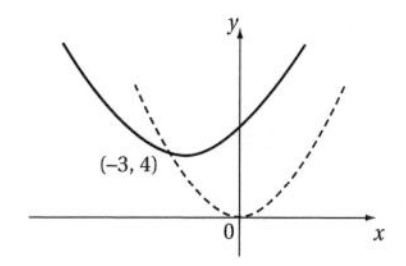

Note:

a Stretch of $y = x^2$, scale factor 2 along the y-axis from 0.

b Stretch of $y = x^2$, scale factor 3 along the y-axis from 0 and reflection in x-axis.

c Stretch of $y = x^2$, scale factor 4 along the y-axis from 0 followed by a translation through $\begin{pmatrix} 0 \\ -1 \end{pmatrix}$.

d Stretch of $y = x^2$, scale factor 2 along the y-axis from 0 followed by a translation through $\begin{pmatrix} -1 \\ 0 \end{pmatrix}$.

e Stretch of $y = x^2$, scale factor 4 along the y-axis from 0 followed by a reflection in the x-axis and a translation through $\begin{pmatrix} 2 \\ 0 \end{pmatrix}$.

f Stretch of $y = x^2$, scale factor $\frac{1}{2}$ along the y-axis from 0 followed by a translation through $\begin{pmatrix} -3 \\ 0 \end{pmatrix}$ and then a translation through $\begin{pmatrix} 0 \\ 4 \end{pmatrix}$.

9 a Translation of $y = \sin x$ through $\begin{pmatrix} 0 \\ 1 \end{pmatrix}$.

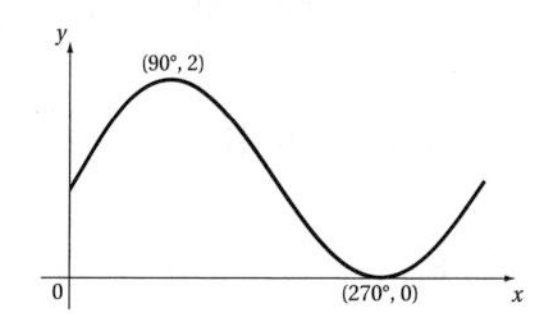

b

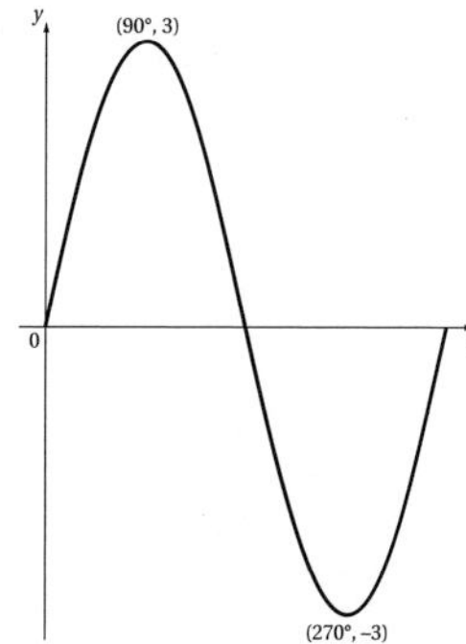

Stretch of $y = \sin x$ scale factor 3 along the y-axis from the origin.

c Stretch of $y = \sin x$ scale factor $\frac{1}{2}$ along the x-axis from the origin.

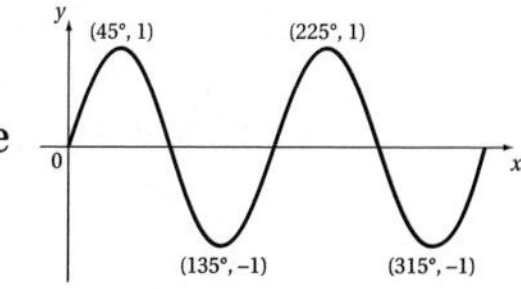

10 a $(180°, -1)$

b i Translation of $y = \cos x$ by $\begin{pmatrix} -90° \\ 0 \end{pmatrix}$.

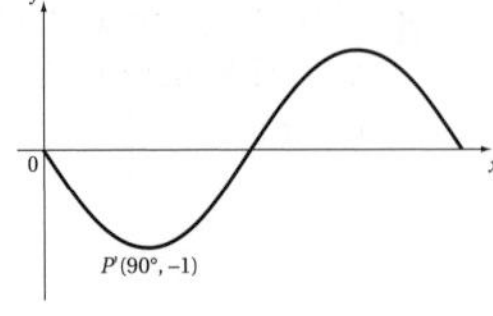

ii Reflection in x-axis.

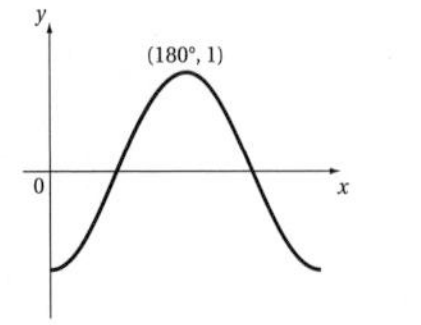

iii Stretch of $y = \cos x$ scale factor 2 along the x-axis from the origin.

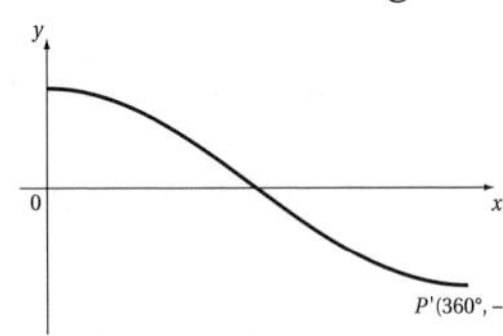

11 a i Stretch of $y = \sin x$ scale factor 2 along the y-axis from 0 followed by a translation through $\begin{pmatrix} 0 \\ -1 \end{pmatrix}$.

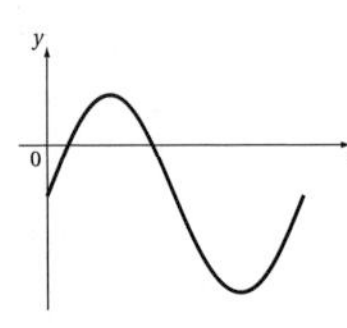

ii Stretch of $y = \sin x$ scale factor 3 along the y-axis from 0 and a reflection in the x-axis.

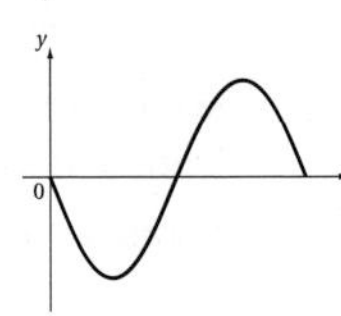

iii Stretch of $y = \sin x$ scale factor 3 along the y-axis from 0 followed by a translation through $\begin{pmatrix} 90° \\ 0 \end{pmatrix}$.

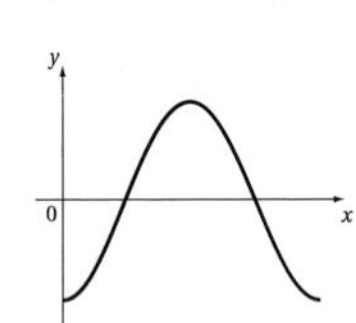

b Graph **iii**, $k = -3$

12 a

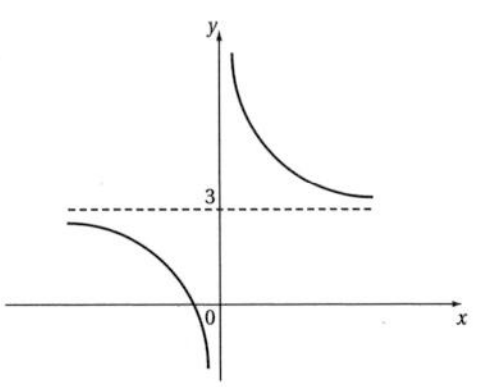

b

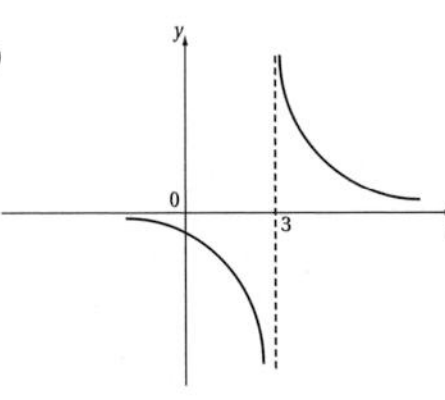

c

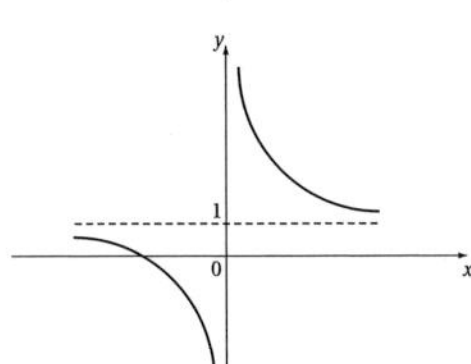

d

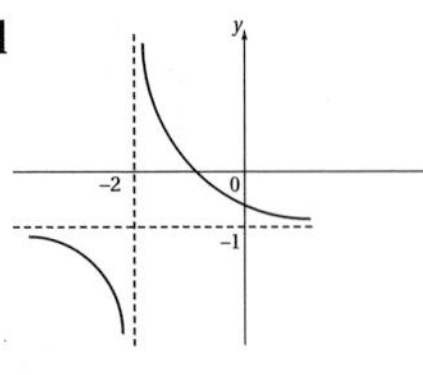

e

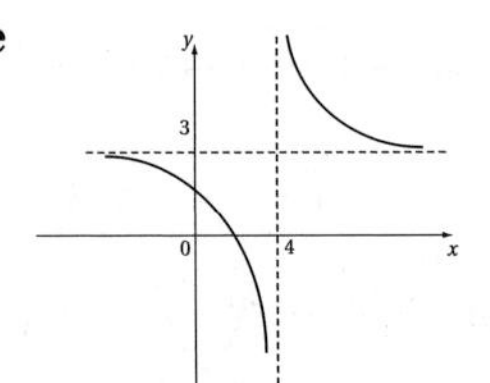

f

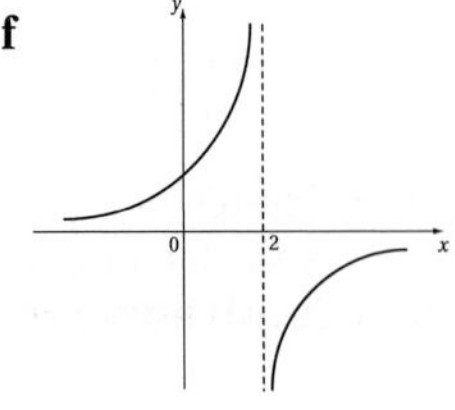

Note:

a Translation of $y = \frac{1}{x}$ through $\begin{pmatrix} 0 \\ 3 \end{pmatrix}$.

b Translation of $y = \frac{1}{x}$ through $\begin{pmatrix} 3 \\ 0 \end{pmatrix}$.

c Stretch of $y = \frac{1}{x}$ scale factor 3 along the y-axis from 0 followed by a translation through $\begin{pmatrix} 0 \\ 1 \end{pmatrix}$.

d Translation of $y = \frac{1}{x}$ through $\begin{pmatrix} -2 \\ -1 \end{pmatrix}$.

e Stretch of $y = \frac{1}{x}$ scale factor 2 along the y-axis from 0 followed by a translation through $\begin{pmatrix} 4 \\ 3 \end{pmatrix}$.

f Stretch of $y = \frac{1}{x}$ scale factor 3 along the y-axis from 0 followed by a translation through $\begin{pmatrix} 2 \\ 0 \end{pmatrix}$ and a reflection in the x-axis.

13 a 5 **b** $y = -1$

c i $\left(0, -\frac{4}{5}\right)$ **ii** $(-4,0)$

14 a $x = 0, y = 1$ **b** $x = 0, y = -2$

c $x = 1, y = 1$

5.3 Intersection points of graphs

1 a $(2,0), (4,0)$ **b** $(1 - \sqrt{2},0), (1 + \sqrt{2},0)$

2 a $A(2,5), B(6,13)$

b Hint: Simplify $\sqrt{(6 - 2)^2 + (13 - 5)^2}$

3 a $(-2,6), (4,0)$ **b** $(0,7), (-3,1)$

c $(-1 + \sqrt{2}, 4\sqrt{2}), (-1 - \sqrt{2}, -4\sqrt{2})$

d $(3 + \sqrt{3}, 2 + \sqrt{3}), (3 - \sqrt{3}, 2 - \sqrt{3})$

4 a $A(1,0), B(6,0)$ **b** $C(2,4), D(5,4)$ **c** 16

5 a $A(0,5), B(0,-13)$

b Hint: Solve $2x^2 - 4x + 5 = 8x - 13$. $(3,11)$

c Hint: The base is the distance AB and the height is the x-coordinate of P.

6 a Hint: Show that solving $x^2 - 4x + 7 = 1 - 2x$ leads to a quadratic equation with negative discriminant.

b The line and curve do not intersect.

c $(x - 2)^2 + 3$ **d**

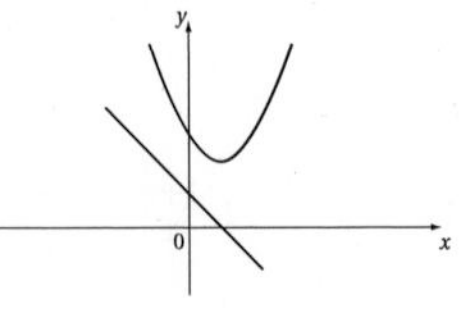

7 a $(x+3)^2+4$ **b**

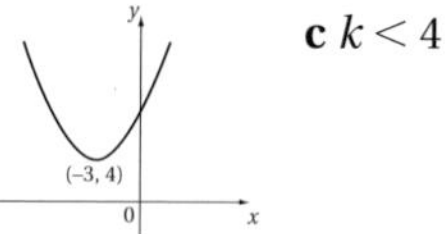

c $k < 4$

8 a Hint: Any solution to $x^2 - 8x + 29 - k = 0$ corresponds to an intersection point of the curve $y = x^2 - 10x + 29$ and line $y = k - 2x$.

b Hint: $x^2 - 8x + 29 - k = (x-4)^2 + 13 - k$.

c (4,5)

9 a Hint: Rearrange $x^2 - 8x + 21 = -x^2 + 6x + 9$.

b $A(1,14)$, $B(6,9)$

c Hint: Start by finding the gradient and the mid-point of AB.

d (−0.2,7.8), (5.2,13.2)

10 a $A(1,8)$, $B(7,20)$ **b** 36

5.4 Three circle theorems

1 a Hint: Use Pythagoras' theorem on the right-angled triangle QPC.

b M lies outside the circle.

2 a Hint: Use Pythagoras' theorem on the right-angled triangle APB.

b 96 cm^2

3 a The perpendicular bisector of AB is the line $x = 5$ and this line must pass through the centre of the circle.

b $\sqrt{7}$

4 a Hint: Use the formula Gradient $= \dfrac{y_2 - y_1}{x_2 - x_1}$

b $\dfrac{1}{2}$ **c** $2y = x + 16$

5 a $-\dfrac{1}{3}$ **b** 3

c $y = 3x - 20$

6 a Hint: Start by calculating the gradient of PC.

b $2y + 3x = 36$

c $A(0,18)$, $B(12,0)$

d Hint: Start by finding the radius r of the circle by calculating the distance CP. The area of the circle is πr^2.

7 a Hint: Start by using the gradient of L_1 to work out the gradient of L_2.

b (3,4)

c Hint: Use the diameter as the base of the triangle.

d 40

6 Answers: Trigonometry and triangles

6.1 Trigonometry and triangles

1 a 6 cm **b** 10 cm **c** 2 cm

2 a Hint: Use $\sin x = \dfrac{\text{opposite}}{\text{hypotenuse}}$

b Hint: Use $AB^2 = AC^2 - BC^2$.

c i $\dfrac{\sqrt{3}}{2}$ **ii** $\sqrt{3}$

3 a 4 **b** $\sqrt{8}$ cm **c i** 1 **ii** $\dfrac{\sqrt{2}}{2}$

4 a $a = \dfrac{b \sin \hat{A}}{\sin \hat{B}}$ **b** $\sin \hat{C} = \dfrac{c \sin \hat{A}}{a}$ **c** $\sin \hat{B} = \dfrac{b \sin \hat{C}}{c}$

5 a 4.43 cm **b** 11.4 cm **c** 11.6 cm **d** 2.99 cm

6 a Hint: Use the sine rule $\dfrac{\sin \hat{B}}{b} = \dfrac{\sin \hat{A}}{a}$. **b** 9.82 cm

7 a Hint: Rearrange $\frac{\sin \hat{B}}{2x} = \frac{\sin 30}{x}$ to make $\sin \hat{B}$ the subject.

b Hint: Find $\sin^{-1}(1)$. **c** $\sqrt{3}x$

8 a $\cos \hat{C} = \frac{a^2 + b^2 - c^2}{2ab}$

b $\hat{B} = \cos^{-1}\left(\frac{a^2 + c^2 - b^2}{2ac}\right)$

9 a 2.97 cm **b** 12.4 cm

10 a $\hat{C} = 83.2°$ **b** $\hat{A} = 53.8°$, $\hat{B} = 43.0°$

11 a Hint: Use the cosine rule $a^2 = b^2 + c^2 - 2bc \cos \hat{A}$.

b 120°

c $\frac{3\sqrt{3}}{4}x^2$ (Hint: The perpendicular from *B* bisects the base *AC*.)

12 a Hint: Use the cosine rule on triangle *BAD*.

b 21.2° **c** 83.8°

13 a Hint: Use the cosine rule on triangle *BCD*, where $BC = 2x$.

b 11.1 cm

14 a Hint: Use the sine rule on triangle *ADB*.

b 93.0°

c Opposite angles in a cyclic quadrilateral sum to 180°.

Since $\hat{A} + \hat{C} = 134.1° + 93.0° = 227.1$, *ABCD* cannot be a cyclic quadrilateral.

Hence all four points cannot lie on a common circle.

d Hint: Start by finding angle *CDB* and then use the cosine rule on triangle *ADC*.

6.2 The area of any triangle

Where appropriate, answers are given to 3 significant figures unless stated otherwise.

1 a 29.7 cm² **b** 27.2 cm² **c** 17.2 cm²

2 23.3 cm²

3 a Hint: Rearrange the sine rule $\frac{\sin \hat{R}}{10} = \frac{\sin 150°}{30}$ to make $\sin \hat{R}$ the subject.

b 9.59° **c** Hint: Use the formula $\frac{1}{2}qr \sin \hat{P}$.

4 10.8 cm²

5 a Hint: Rearrange $c^2 = a^2 + b^2 - 2ab \cos \hat{C}$ to make $\cos \hat{C}$ the subject.

b 26.8 cm²

6 a 64π cm² **b** 22.6 cm²

c Hint: The sector *CAB* has area $\frac{1}{8} \times \pi(8)^2$.

7 a 9.00 cm² **b** 3.01 cm

c Hint: Start by using the cosine rule to find the length of the side *PR*.

8 a Hint: Simplify $\frac{1}{2}(x)(2x) \sin 30°$.

b 6 cm

c Hint: Use the cosine rule $p^2 = q^2 + r^2 - 2qr \cos \hat{P}$.

d 4.84 cm

9 a 49.8 cm²

b Hint: the sector has area $\frac{17}{72} \times \pi\ (10)^2$.

10 a Hint: Start by finding angle *BCA* and then use the sine rule.

b 47.3 cm² **c** 35.6 cm

11 a Hint: Start by finding the length *AE* by using Pythagoras' theorem on triangle *ABE*. Then use the cosine rule on triangle *AEF*.

b $\frac{5}{2}$ cm²

c Hint: Start by calculating angle *BEA* using right-angled trigonometry. Then find angle *CEF*.

d $\frac{104}{25}$ cm²

6.3 Solving a trigonometric equation

Where appropriate, answers are given to 1 decimal place.

1 a $x = 0°, x = 180°, x = 360°$ **b** 36.9° **c** 143.1°

2 a 180° **b** $x = 134.4°, x = 225.6°$

3 a 90° **b** $x = 14.5°, x = 165.5°$

c $x = 30°, x = 150°, x = 390°, x = 510°$

4 a $x = 0°, x = 360°$ **b** $x = 48.2°, x = 311.8°$

c $x = 135°, x = 225°, x = 495°, x = 585°$

5 a Hint: Both answers are equal to $\frac{1}{\sqrt{2}}$.

b

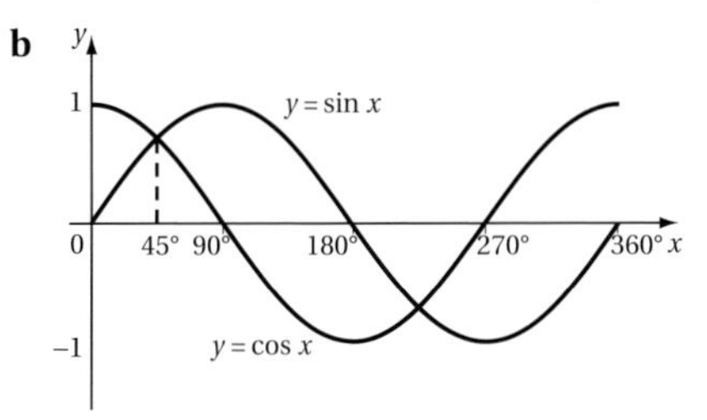

c 45° (see sketch) **d** 225°

6 a Hint: Rearrange the sine rule $\frac{\sin \hat{B}}{b} = \frac{\sin \hat{A}}{a}$ to make $\sin \hat{B}$ the subject.

b 34.8° **c** 145.2° **d** e.g.

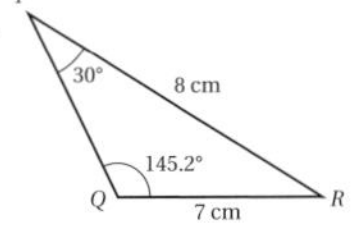

7 a $x = 30°, x = 150°$ **b** $x = 70.5°, x = 289.5°$

c $x = 45°, x = 135°$ **d** $x = 75.5°, x = 284.5°$

8 a Hint: Apply a suitable translation to the graph of $y = \sin x$.

b $x = 78.5°, x = 281.5°$

c $x = 48.6°, x = 131.4°$

9 a $-30°$ **b** 210°

c $x = 233.1°, x = 306.9°$

10 a Hint: Sketch the graph of $y = \sin x$ and the line $y = \frac{5}{4}$ on the same diagram.

b i Not possible as $\cos x \leqslant 1$ for all x.

ii 270°

iii Not possible as $\sin x \geqslant -1$ for all x.

iv $x = 41.4°, x = 318.6°$

11 a Stretch, scale factor $\frac{1}{a}$, along the x-axis from the origin.

b $x = 23.6°, x = 156.4°$

c i $x = 11.8°, x = 78.2°$ **ii** $x = 5.9°, x = 39.1°$

iii $x = 15.7°, x = 104.3°$

12 a $x = 112.0°, x = 248.0°$

b i $x = 37.3°, x = 82.7°$ **ii** $x = 56.0°, x = 124.0°$

13 a $x = -286.3°, x = -73.7°, x = 73.7°, x = 286.3°$

b $x = -319.5°, x = -220.5°, x = 40.5°, x = 139.5°$

c $x = -135.6°, x = -44.4°, x = 224.4°, x = 315.6°$

d $x = -246.4°, x = -113.6°, x = 113.6°, x = 246.4°$

7 Answers: Sequences and summations

7.1 Arithmetic sequences

1 a i $3n + 2$ **ii** $2n - 7$ **iii** $-4n + 16$

iv $\frac{1}{2}n + \frac{1}{2}$ **v** $-\frac{3}{2}n + 7$ **vi** $\frac{5}{4}n - \frac{1}{3}$

b i 32 **ii** 13 **iii** -24 **iv** $\frac{11}{2}$

v -8 **vi** $\frac{73}{6}$

2 a Not arithmetic (differences are $-3, -3, 3$ …)

b Could be arithmetic, with common difference 9

c Not arithmetic (differences are $-\frac{1}{2}, -\frac{1}{6}$ …)

d Could be arithmetic, with common difference 0

e Could be arithmetic, with common difference $\sqrt{2}$

f Not arithmetic (differences are 2, 2, 4…)

3 a $4n + 3$ **b** 83 **c** The 42nd term

d Because 98 is not divisible by 4

4 a 12 **b** $3n + 6$ **c** Because $3n + 6 = 3(n + 2)$

d The 65th term

5 a $4n + 6$ **b** 134

c Hint: Show that $u_{n+1} = 4n + 10$ and then simplify $(u_n + u_{n+1})$.

6 a $a = 4, b = 18$ **b** $7n - 3$

c $u_1 = 4, u_4 (= 25), u_{12} (= 81)$

7 a -5 **b** The 8th term **c** 3

8 a $50n + 2000$ **b** £2600

c The 40th month

9 a Hint: Use the formula $u_n = a + (n - 1)d$ where $a = 9000, d = -100$.

b 81 (allow 80) months

c i £11000 **ii** £115 per month

10 a $\frac{1}{2}n + 1$ **b** The 8th day

c 15 hours. This seems excessive, although his exams are now very close.

d 21 hours

11 a $a + 2d = 8$ **b** $a + 13d = 63$

c -2 **d** 243

12 a $u_n = a + (n - 1)d$

b Hint: Express the equation $u_5 = 3u_2$ in terms of a and d.

c $a = 2, d = 4$

d Because $u_n = 4n - 2$ and 2 is not divisible by 4.

13 a Hint: Show that the difference of consecutive terms is -1.

b $4(k + 1)(k - 1)$ **c** $k = \frac{3}{2}, k = -\frac{3}{2}$

7.2 Recurrence relations and sigma notation

1 a 5, 8, 11 **b** 6, 12, 24 **c** 1, 7, 25

d $-1, 5, -1$ **e** $3, -3, 9$ **f** 2, 2, 2

2 a $u_2 = -3, u_3 = 3, u_4 = -3$ **b** 3 **c** -9

3 a $u_2 = 9, u_3 = 12$ **b** Arithmetic

c $3n + 3$ **d** 75

4 a $2k + 7$ **b** Hint: Solve $2k + 7 = 3$.

c 33

5 a $8 + k$

b Hint: Use the relationship $u_3 = 2u_2 + k$.

c -3 **d** 19

6 a $2k + 4$

b Hint: Use the relationship $u_3 = ku_2 + 4$.

c $k = 3, k = -5$ **d** $u_4 = 106, u_4 = -166$

7 a i $a = 3, b = 1$ **ii** 40

b i $a = -2, b = 7$ **ii** 13

c i $a = \frac{1}{2}, b = \frac{3}{2}$ **ii** $\frac{11}{4}$

8 a 26 **b** 0

c 30 **d** 137

9 a $4n - 2$ **b** 18

c e.g. $\sum_{n=1}^{18} (4n - 2)$

10 a $1 + 4 + 9 + 16 = 30$

b $1 + 6 + 15 + 28 = 50$

c $\frac{1}{2} + \frac{3}{2} + \frac{9}{2} = \frac{13}{2}$

d $13 + 7 + 5 + 4 = 29$

e $-2 + 4 - 8 + 16 - 32 = -22$

f $3 + 4 + 5 + 6 = 18$

11 a $-3n + 33$ **b** The 11th term **c** 21

12 a Hint: Work out u_2 and then u_3. **b** 157

13 a Hint: Use the relationship $u_2 = au_1 - 1$

b 182

8 Answers: Introducing differentiation

8.1 Estimating the gradient of a curve

1 a 5 **b** 3 **c** 4

2 a i 6 underestimate **ii** 10 overestimate

iii 9.66 (3 sig. figs) overestimate

b Answer iii, as $\sqrt{2} \approx 1.41$ so S is the closest of the three points to P.

3 a $(x-2)^2+3$ **b** (2,3) **c** 0

d Hint: This follows from $Grad_{PQ} = 3$, $Grad_{RP} = 1$

e -2

4 a 8 **b** -12 **c** -7 **d** 1

5 e.g. choose (2.9,7.1025) and (3.1,7.4025).

$1.475 < grad < 1.525$ so gradient at $x = 3$ is 1.5 (1 decimal place).

6 e.g. choose (1.99,0.49246…) and (2.01,0.50746…).

$0.746\ldots < grad < 0.753\ldots$ so gradient at $x = 2$ is 0.75 (2 decimal places).

7 a From the diagram, $m < Grad_{PQ} = 5$

b $P(1,2)$, $Q(1.5,4.5)$

c For the lines to intersect at a point with a positive x-coordinate, the tangent must be steeper than the line OQ.

So $m > Grad_{OQ}$ where $Grad_{OQ} = \frac{4.5}{1.5} = 3$

d $m = 4$, $y = 4x - 2$

8.2 Differentiation

1 a Hint: Use the formula $grad = \frac{y_2 - y_1}{x_2 - x_1}$ and then simplify by factorising.

b $6x$

2 a $q^2 - 3q + 4$

b Hint: Use the formula $grad = \frac{y_2 - y_1}{x_2 - x_1}$ and then simplify by factorising.

c $2x - 3$

3 a $(q-x)(q+x)$ **b** $(q-x)(q+x)(q^2+x^2)$

c Hint: Show that $Grad_{PQ} = (q+x)(q^2+x^2)$ and then replace q with x.

4 a $2x+3$ **b** $10x$ **c** $-6x$

d $x+5$ **e** $4-4x$ **f** $-7-6x$

5 a $2x+5$ **b** $4x-5$ **c** $4-6x$

d $-30x+26$ **e** $8x+12$ **f** $-3+\frac{1}{2}x$

6 a 7 **b** $-\frac{3}{2}$ **c** 16 **d** -5

7 a $x^2(4x+3)$ **b** $3(x+2)(x-2)$

c $3(x-2)^2$ **d** $4x(x+1)(x-1)$

e $x(x+2)(x+4)$ **f** $(5x^2+1)(x^2+1)$

8 a $y = x^3 - 8$ **b i** $\frac{1}{3}$ **ii** 12

9 a $8x-3$ **b** $\frac{3}{8}$ **c** $\left(\frac{1}{2},\frac{3}{2}\right)$

10 a (2,7) **b** $\left(1,\frac{5}{4}\right)$ **c** $(-2,-16)$

11 a Hint: Find $\frac{dy}{dx}$ and then substitute in the value $x = 1$.

b $(-3,2)$

12 a Hint: Start by showing that when $x = 2$, $\frac{dy}{dx} = 4k$.

b $\frac{3}{2}$

c The gradient of the curve at B must be between 1 and $\frac{3}{2}$, but $\frac{7}{4} > \frac{3}{2}$.

13 a $3x^2$ **b** $6x^2-3$ **c** $4x-\frac{9}{2}$

d 1 **e** $\frac{2}{3}$ **f** $2x$

14 a Hint: Start by showing that $\frac{1}{q} - \frac{1}{x} = \frac{-(q-x)}{qx}$.

b $-\frac{1}{x^2}$

c Hint: Write the equation of the curve $y = \frac{1}{x}$ as $y = x^{-1}$ and then use the nx^{n-1} rule and the rules of indices.

15 a $q - x$

b Hint: Factorise the numerator of $Grad_{PQ}$ using the result of part **a**.

c $\dfrac{1}{2\sqrt{x}}$

d Hint: Write the equation of the curve $y = \sqrt{x}$ as $y = x^{\frac{1}{2}}$ and then use the nx^{n-1} rule and the rules of indices.

8.3 Applications of differentiation

1 a Hint: Find an expression for $\frac{dy}{dx}$ and substitute in the value $x = 2$.

b $y = 3x - 2$

2 a $y = -4x + 2$ **b** $y = 9x - 17$

c $y = 2x - 1$ **d** $y = -5x + 8$

3 $y = -6x + 8$

4 a Hint: Start by substituting the value $x = \frac{1}{2}$ into the equation and also into $\frac{dy}{dx}$.

b $4y + 16x = 3$

5 a 2

b Hint: Use $Grad_N \times Grad_T = -1$.

6 a $5y - x - 13 = 0$ **b** $2y - x - 5 = 0$

c $4y + x + 78 = 0$

7 a (0,7)

b Hint: Start by substituting the value $x = 2$ into the equation and also into $\frac{dy}{dx}$.

c 12

8 a Hint: Start by substituting the value $x = 1$ into $\frac{dy}{dx}$

b $y = \frac{x}{4} + \frac{11}{4}$

c Hint: Use the tangent and normal equations to find the coordinates of A and B respectively.

d i $\frac{17}{8}$ **ii** $\frac{17}{4}$

9 a Hint: The tangent has gradient 2, so $\frac{dy}{dx} = 2$ when $x = 3$.

b $y = -10x - 21$ **c** $(2,-16)$

10 a Hint: Start by finding the gradient of the curve at point P.

b (1,6)

11 a i $(3,-5)$ **ii** $\left(-\frac{1}{2},6\right)$

b i

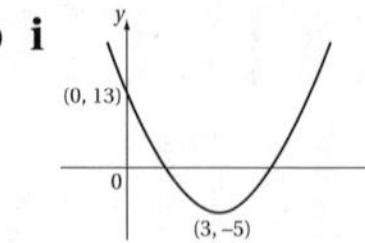

ii

(−½, 6)
(0,5)

12 a Hint: Show that $\frac{dy}{dx} = 0$ leads to the equation $x^2 + x - 2 = 0$.

b $P(1,-2)$, $Q(-2,25)$

13 a i $(2,-6)$, $(-2,26)$ **ii** $(3,-2)$, $(1,2)$

iii $(0,7)$, $(2,11)$ **iv** $(0,3)$, $(3,-24)$

v $(1,5)$, $(-1,1)$ **vi** $\left(\frac{1}{2},-\frac{11}{4}\right)$, $(1,-3)$

b i

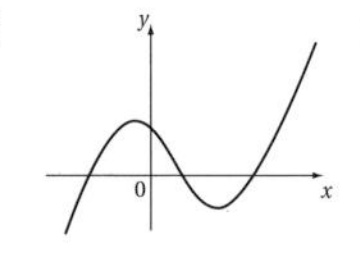

ii

iii

iv

v

vi

14 a Hint: Find $\frac{dy}{dx}$ and then show the equation $\frac{dy}{dx} = 0$ has no real solutions.

b Hint: Find $\frac{dy}{dx}$ and then use the quadratic formula on the equation $\frac{dy}{dx} = 0$.

c Hint: Find $\frac{dy}{dx}$ and then use the quadratic formula on the equation $\frac{dy}{dx} = 0$, leaving the answers in terms of k.

15 a Hint: sketch a cubic graph for extreme values (positive or negative) of x.

b If $y = ax^4 + bx^3 + cx^2 + dx + e$ then $\frac{dy}{dx} = 4ax^3 + 3bx^2 + 2cx + d$.

The equation $\frac{dy}{dx} = 0$ is a cubic equation and so by James' statement is guaranteed to have a solution.

c Tom is not correct. e.g. the curve $y = x^3 + x$ has no stationary points since $\frac{dy}{dx} = 3x^2 + 1 > 0$ for all x.

Exam paper

Total = 50 marks

Time allowed: 1 hour

1 a Find the value of $27^{-\frac{1}{3}}$. **[2]**

b Express $\frac{12}{3 - \sqrt{5}}$ in the form $p + q\sqrt{5}$, where p and q are integers. **[3]**

---

2 The points A and B have coordinates $A(2,4)$ and $B(-8,9)$.

a Find the distance AB. Give your answer in simplified surd form. **[3]**

b Find an equation for the line which passes through $A(2,4)$ and $B(-8,9)$.
Give your answer in the form $ay + bx = c$ where a, b and c are integers. **[4]**

3 The diagram shows the graph of $y = f(x)$. The graph crosses the x and y-axes at the points $A(4,0)$ and $B(0,6)$, respectively. Point $C(2,7)$ is a maximum point on this graph.

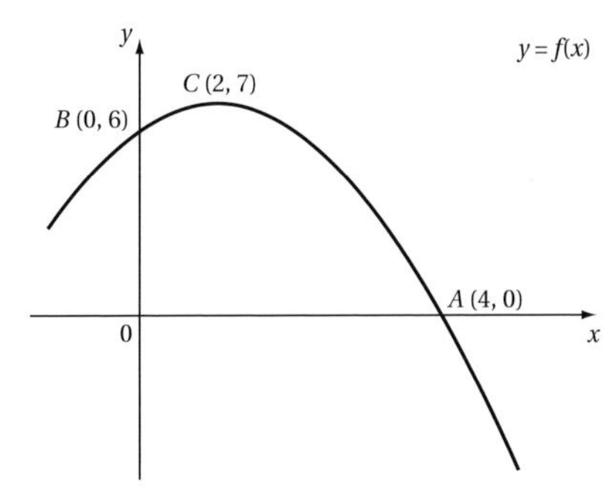

On the diagrams provided, sketch the graph with equation

a $y = f(2x)$ **[2]**

b $y = f(-x)$ **[2]**

On each diagram, show the coordinates of the points corresponding to A, B and C.

a

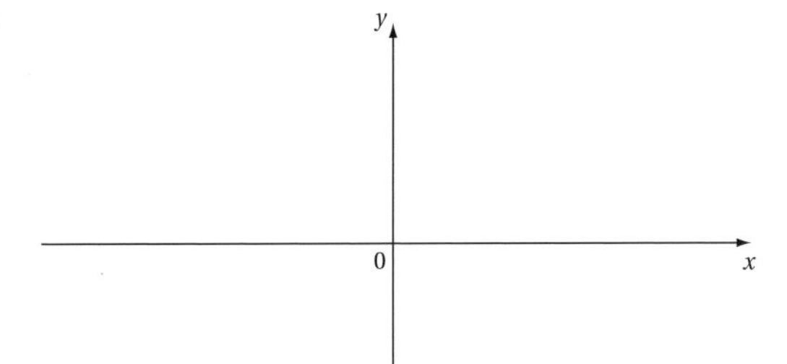

b

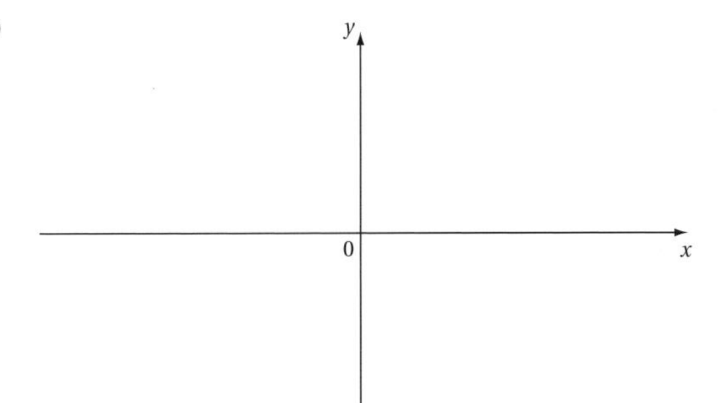

4 Find the set of values of x for which

a $\frac{7 - 4x}{3} \leqslant 1$ **[2]**

b $-4 \leqslant 3x + 5 < 11$ **[2]**

c **both** $\frac{7 - 4x}{3} \leqslant 1$ **and** $-4 \leqslant 3x + 5 < 11$ **[1]**

5 a Express $x^2 - 6x + 13$ in the form $(x - p)^2 + q$, where p and q are constants. **[2]**

b Use your answer to part **a** to describe the single transformation which maps the graph of $y = x^2$ onto the graph of $y = x^2 - 6x + 13$. **[2]**

c In the space provided, sketch the graph of $y = x^2 - 6x + 13$, clearly marking the coordinates of the minimum point. **[2]**

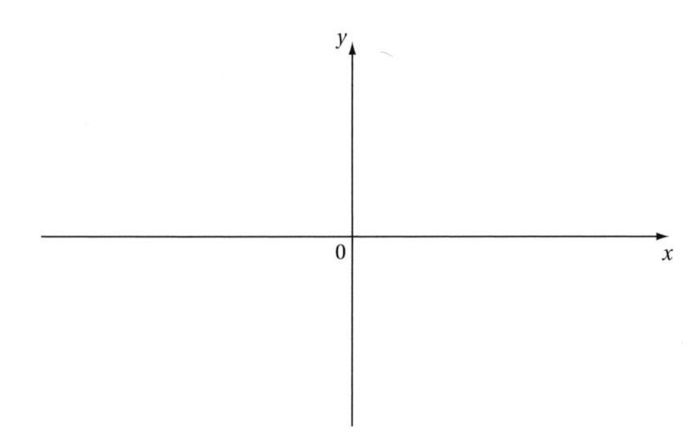

d Express, as an inequality, the range of values of k for which the equation $x^2 - 6x + 13 - k = 0$ has no real solutions. **[2]**

6 The shape in the diagram shows part of a compact disc storage rack in which triangle ABC has been joined to the sector ACD.

The sector of the circle has centre A and radius r cm. $AB = 3$ cm, $BC = 7$cm and angle $BAC = 60°$.

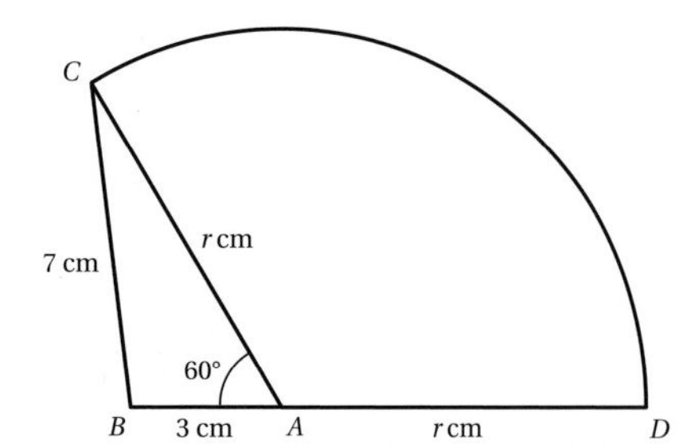

It is given that $\cos 60° = \frac{1}{2}$.

a Use the cosine rule on triangle ABC to show that r satisfies the equation $r^2 - 3r - 40 = 0$. **[3]**

b Hence find the radius of the sector. **[2]**

Given also that $\sin 60° = \frac{\sqrt{3}}{2}$,

c show that the area of the shape is $6\sqrt{3} + \frac{64}{3}\pi$ cm^2 **[4]**

7 The diagram shows the graph of $y = x^2 - 8x + 15$. The graph crosses the x-axis at the points A and B. Point P is the minimum point on the graph. The vertical dotted line through P is a line of symmetry of this graph.

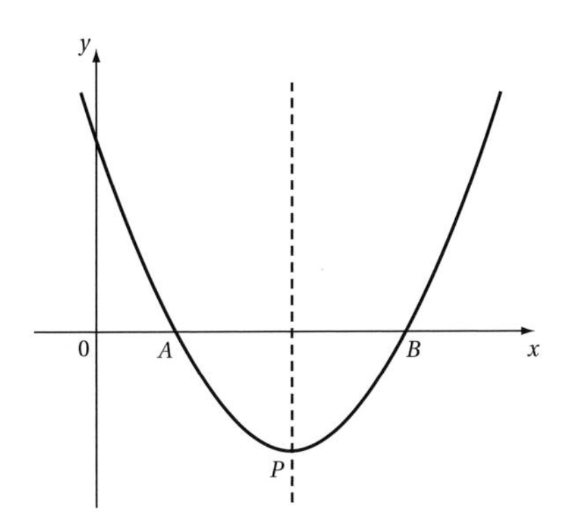

a Find the coordinates of A and B. **[3]**

b Hence, or otherwise, find the coordinates of P. **[2]**

c Show that the line L perpendicular to AP and which passes through the mid-point of AP has equation $y = x - 4$. **[4]**

[*continued*]

d Find the x-coordinates of the points where L intersects the graph of $y = x^2 - 8x + 15$.

Give your answers in surd form. **[3]**

EXAM PAPER